Biocatalytic Membrane Reactors

Applications in biotechnology and the pharmaceutical industry

ENRICO DRIOLI
LIDIETTA GIORNO

Research Institute on Membranes and Modeling of Chemical Reactors, CNR and Department of Chemical Engineering and Materials, University of Calabria, Italy

CRC Press
Taylor & Francis Group
Boca Raton London New York

CRC Press is an imprint of the
Taylor & Francis Group, an **informa** business

A TAYLOR & FRANCIS BOOK

CRC Press
Taylor & Francis Group
6000 Broken Sound Parkway NW, Suite 300
Boca Raton, FL 33487-2742

© 1999 by Taylor & Francis Group, LLC
CRC Press is an imprint of Taylor & Francis Group, an Informa business

Visit the Taylor & Francis Web site at
http://www.taylorandfrancis.com

and the CRC Press Web site at
http://www.crcpress.com

Contents

Preface

This book provides an analysis of biocatalytic membrane reactors. It includes basic aspects and applications of catalytic membrane systems on topics of current interest, particularly in the biotechnology and pharmaceutical fields. Recent progress and potential developments of membrane (bio)technology are presented.

The book consists of two parts. The first introduces general principles of membrane processes (Chapter 1) and catalytic membrane reactors (Chapter 2). The second is concerned with practical applications of biocatalytic membrane reactors. In Chapter 3, their use in the processing of bioactive compounds produced by fermentation is discussed. The concepts of the continuous membrane fermentor and integrated membrane processes are described using as examples the production of organic acids. Some examples of the use of integrated membrane systems for the processing of food and beverages are also given.

In Chapter 4, catalytic membrane reactors and membrane systems for obtaining optically pure enantiomers are presented.

The particular case of catalytic membrane reactors using cofactor dependent enzyme, where the retention and regeneration of cofactor is needed, is discussed in Chapter 5.

In Chapter 6, the use of membrane reactors in biphasic organic/aqueous systems and in pure organic solvents is presented.

The use of membrane reactors for biomedical applications is discussed in Chapter 7.

Enrico Drioli
Lidietta Giorno

Introduction

Membrane science and technology in the last twenty years have had an impressive growth, confirming their potential to contribute to the solution of crucial problems of the world today and to sustainable industrial development. Molecular separation processes based on polymeric, ceramic and liquid membranes have been studied and applied in probably a wider spectrum of areas than has any other existing technology. The continuing progress in the understanding of the basic principles controlling their selectivity and permeability, and the efforts for better design and engineering of the various membrane operations already developed, will contribute to the rationalization of different industrial applications, to the creation of new artificial organs and to their integration, and to the exploitation of new frontiers (e.g. the transformation of light signals into electrical signals). The fact that membranes can be used not only for molecular separations but also for carrying out reactions, in the same way as living organism achieve their biochemical transformations, has attracted and is attracting the interest of scientists and engineers.

Catalytic membrane reactors are interesting with respect to conventional ones owing to the advantages of combining selective mass transfers across the membrane with chemical reactions and because they represent, in principle, ideal surfaces for reactions.

Provided membranes of suitable molecular mass cut-off are used, chemical reactions and physical separations of catalysts (and/or reagents) from the product may take place in the same unit.

Selective removal of products through membranes yields effective conversions with product-inhibited or thermodynamically unfavourable reactions. Because of the in-situ separation of chemical species, reactions are displaced from equilibrium conditions, thus yielding faster and higher conversions, possibly at lower temperatures and lower costs.

Owing to their large available surface per unit volume and offering new possibilities for various immobilization procedures, synthetic polymeric or ceramic membranes provide ideal supports for catalyst and biocatalyst immobilization. The key attribute of membranes that makes them particularly attractive for bioreactor applications is their ability to compartmentalize. Immobilized enzymes are retained in the reaction space and can be reused continuously. Immobilization has also been proven to enhance enzyme stability (Drioli et al., 1975; Messing, 1975).

Membrane reactors can also be realized by integrating traditional chemical reactors (e.g. CSTR) with one or more membrane operations (e.g. ultrafiltration, pervaporation, dialysis, etc.). The flexibility of the membrane separation units — their ability to operate in gas or liquid phase, using different driving forces (ΔC, ΔP, ΔT, etc.) or appropriate carriers — can be exploited for the design and control of the reactions.

Catalytic membranes and membrane reactors have been studied until recently mainly for bioconversions. The opportunity of increasing enzyme lifetime by immobilization, retaining expensive cofactors (often required in bioconversion) in the reaction volume, removing enzyme inhibitors continuously, and controlling sequential reactions, provides interesting possibilities with these systems.

Enzymatic membranes will also contribute to the growth of new areas of industrial biochemistry, such as nonaqueous enzymology. It has been shown that most enzymes are able to operate in organic solvents containing no or little water (Zaks and Klibanov, 1988). It has also been shown that the change in solvents dramatically affects key catalytic characteristic of enzymes, including thermostability, enantioselectivity, substrate specificity and regioselectivity. The use of organic solvents as reaction media is of particular interest because (a) various substrates have a low solubility in water; (b) some side reactions such as hydrolysis or isomerization can be avoided in nonaqueous media; and (c) microbial contamination can be minimized. Higher enzyme concentration might be necessary, however, to increase reaction rates, which are often slower in organic solvents. Immobilized enzymes seem to show higher activity with respect to the enzymes in solution, when used in organic solvents.

The possibility of immobilizing nonbiological catalysts such as cyclodextrins, for example, in polymeric membranes made by phase inversion has been also discussed and tested (Drioli et al., 1995). The specific selectivity of these chemical species, their stability in organic media and their potential catalytic activities seem to be well utilized in membrane configurations.

When cells are so big as to prevent their entrapment inside the membrane wall, membranes may still be used to confine them in a region of the membrane module, such as the module shell-side. In the design of a bioartificial membrane pancreas, islets of Langerhans (the cellular aggregates responsible for pancreatic glucose-dependent insulin secretion) were confined in the shell of a membrane module equipped with hollow-fibre membranes arranged in 'tube-and-shell' configuration. The size of islets (about 150 µm diameter) and their susceptibility to mechanical and chemical stresses prevented the adoption of more effective but more damaging immobilization techniques. Membranes were then used to 'immobilize' the biocatalytic principles in a rather mild way. In this case, however, they were mainly chosen to operate as selective barriers to prevent any contact between allo- or xenograft (from a different source than the recipient) islets of Langerhans and the constituents of the immune system of the recipient whose blood is circulated in the membrane core. Membranes have to ensure free permeation of nutrients and glucose from the bloodstream to the islets and of catabolites and secreted hormones back to the bloodstream, while preventing rejection of the implanted foreign tissue (Gerlach, 1996).

A new class of membranes made of inorganic compounds has been recently developed. Inorganic membranes (dense or porous), do not have the limitations associated with the traditional polymeric ones: they have a high degree of resistance to chemical and abrasive degradation, and tolerate a wide range of pH and temperature (Hsieh, 1996). Inorganic membrane reactor technology is today considered a generic alternative to conventional reactor design, particularly for dehydrogenations, hydrogenations, oxidations, and the like

(Kikuchi, 1997). The potential of these new membranes in pharmaceutical and bio-technological applications is also interesting. The integration of the available membrane separation processes and membrane reactors with traditional unit operation permits the redesign of innovative production lines, particularly important in bioproduction.

In conclusion, every catalytic industrial process can potentially benefit from the intro-duction of catalytic membranes and membrane reactors in place of the conventional catalytic reaction systems: because of the *in situ* separations of chemical species, and of the potential for providing higher conversions and yields, minimizing catalyst deactiva-tion, controlling side reactions and optimizing recovery and reuse of reagents and by-products. Their conceptual simplicity, the facility in scaling up, and the possibility of total control and automatization will also contribute to successful industrial realization of these systems.

The biotechnology industry, the fastest-growing sector of the chemical process industry, might particularly benefit from these new technologies. In 1996, 1290 biotechnology firms in the United States realized sales of US$10.8 billion and revenues of US$14.6 billion, up 15% on 1995 levels. An annual business growth of 11% over the next five years is expected. In Europe this industrial sector is also in expansion. In September 1996, EUROBIO was created to represent over 500 biotech firms. In the United Kingdom and in Germany, significant public funds are today devoted to supporting companies in this area.

At the 1997 ACHEMA in Frankfurt, statements on the impact that innovative bioreactors, and particularly those based on hollow-fibre designs, have in setting new performance standards were clearly presented. Hollow-fibre bioreactors, for example, in which cells attach into a capillary-type space, are designed to mimic biological processes more closely than any other reactor design (tank; fluidized-bed, etc.). Through the fibres, nutrients such as glucose and oxygen are fed to the cells and wastes such as CO_2 and H_2O are removed. Roche Diagnostic (a subsidiary of Hoffmann-La Roche) reported the use of such reactors to produce monoclonal antibodies for diagnostic products. Researchers culture mouse hybridoma cells, which they inoculate into a hollow-fibre reactor. Yield is about 50 mg l^{-1} of antibody.

A set-up integrating five five-cassette bioreactors, as shown in Figure I.1, allowed the production of 10 to 20 different hybridoma cells at the same time at the Swiss Institute for Cancer Research, doubling the normal production rate.

The overall commercial relevance of membrane processes today is already significant, with sales of membranes and modules in 1994 being over US$3500 million and higher than US$10 billion for the sales of membrane systems. An average growth of 8–10% per year is expected, based on stricter environmental laws and higher energy and raw materials costs. The biomedical market has so far been the largest market.

The most relevant technical applications of synthetic membranes are summarized in Table I.1.

Recent developments have been, in particular, in integrated membrane processes, in membrane reactors and catalytic membranes, and in the immunoisolation of cell cultures. Electrodialytic dissociation of water and new artificial organs such as artificial liver and pancreas are other examples of projects in progress. In Figure I.2 a systematic description of an integrated membrane system in the dairy industry is presented. Other similar schemes are becoming practicable in various biotechnological and pharmaceutical applications. Membrane reactors might be considered one of the most recent successes of membrane research and in large part still in their infancy. Some examples of membrane reactors studied, and in some cases also realized, in pharmaceutical, biomedical and agro-food applications are shown in Tables I.2, I.3 and I.4 respectively.

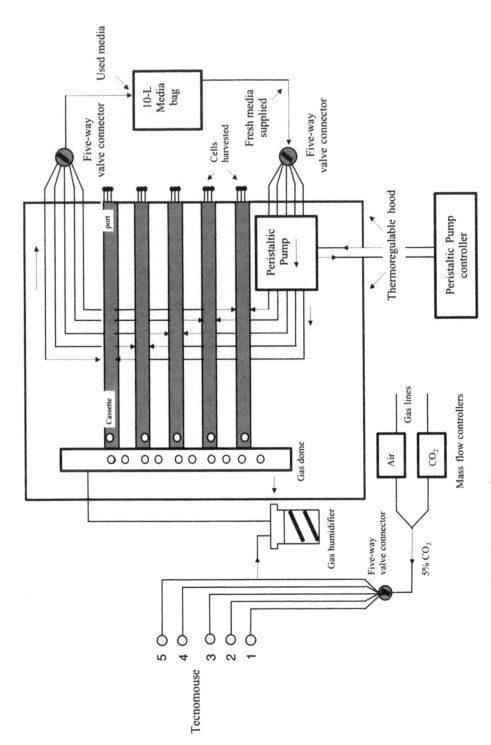

Figure I.1 A set-up integrating a five-cassette bioreactor.

Table I.1 List of the most relevant technical applications of synthetic membranes

Membranes	Basic materials	Manufacturing procedures	Structures	Applications
Ceramic membranes	Clay, silicate, aluminum oxide, graphite, metal powder	Pressing and sintering of fine powders	Pores from 0.1 to 10 μm diameter	Filtering of suspensions, gas separations, separation of isotopes
Stretched membranes	Polytetrafluoroethylene, polyethylene, polypropylene	Stretching of partially crystalline foil perpendicular to the orientation of crystallites	Pores of 0.1–1 μm diameter	Filtration of aggressive media, cleaning of air, sterile filtration, medical technology
Etched polymer films	Polycarbonate	Irradiation of a foil and subsequent acid etching	Pores of 0.5– 10 μm diameter	Analytical and medical chemistry, sterile filtration
Homogeneous membranes	Silicone rubber, hydrophobic liquids	Extrusion of homogeneous foils, formation of liquid films	Homogeneous phase, support possible	Gas separations, carrier-mediated transport
Symmetrical microporous membranes	Cellulose derivatives, polyamide, polypropylene	Phase-inversion reaction	Pores of 50– 5000 nm diameter	Sterile filtration, dialysis, membrane distillation
Integral asymmetric membranes	Cellulose derivatives, polyamide, polysulfone, etc.	Phase-inversion reaction	Homogeneous polymer or pores of 1–10 nm diameter	Ultrafiltration, hyperfiltration, gas separations, pervaporation
Composite asymmetric membranes	Cellulose derivatives, polyamide, polysulfone, polydimethylsiloxane	Application of a film to a microporous membrane	Homogeneous polymer or pores of 1–5 nm diameter	Ultrafiltration, reverse osmosis, gas separations, pervaporation
Ion exchange membranes	Polyethylene, polysulfone, poly(vinyl chloride), etc.	Foils from ion exchange resins of sulfonation of homogeneous polymers	Matrix with positive or negative charges	Electrodialysis, electrolysis

Table 1.2 Membrane reactors for pharmaceutical applications

Reaction	Membrane reactor	Purpose
Production of ampicillin and amoxycillin (penicillin amidase)	Entrapment in cellulose triacetate fibers	Production of antibiotics
Conversion of fumaric acid to L-aspartic acid (*E. coli* with aspartase)	Entrapment in polyacrylamide gel	Pharmaceutical and food additive, raw material for aspartame
Conversion of L-aspartic acid to L-alanine (*P. dacunhae*)	Entrapment in polyacrylamide gel	Pharmaceutical additive
Conversion of cortexolone to hydrocortisone and prednisolone (*C. lunata/C. simplex*)	Entrapment in polyacrylamide gel	Production of steroids
Conversion of acetyl-D,L-amino acid to L-aminoacid (aminoacylase)	Ionic binding to DEAE-sephadex	Production of L-amino acid for pharmaceutical use
Synthesis of tyrosine from phenol, pyruvate and ammonia (tyrosinase)	Entrapment in cellulose triacetate membrane	Production of L-amino acid for pharmaceutical use
Hydrolysis of a cyano-ester to ibuprofen (lipase)	Entrapment in biphasic hollow fibre reactor	Production of anti-inflammatories
Hydrolysis of a diltiazem precursor (lipase)	Entrapment in biphasic hollow fibre reactor	Production of calcium channel blocker
Hydrolysis of 5-p-HP-hydantoine to D-p-HP-glycine (hydantoinase/carbamilase)	Entrapment in UF polysulfone membrane	Production of cephalosporin
Dehydrogenation reactions (NAD)(P)H dependant enzyme systems)	Confination with UF-charged membrane	Production of enantiomeric amino acids
Hydrolysis of DNA to olygonucleotides (DNase)	Gelification on UF capillary membrane	Production of pharmaceutical substances

Table I.3 Membrane reactors for biomedical applications

Reaction	Membrane reactor	Purpose
Hydrolysis of hydrogen peroxide (bovine liver catalase)	Entrapment in cellulose tryacetate membrane	Liver failure
Hydrolysis of insulin to glucose and levulose (insulase)	Entrapment in sodium-alginate gel	Care of liver disease
Oxidation of bilirubin (bilirubin oxidase)	Immobilization on sepharose	Prevention of jaundice in newborn
Degradation of heparin (heparinase)	Immobilization on sepharose	Heparin neutralization
Hydrolysis of arginine and asparagine (arginase and asparaginase)	Entrapment in polyuretane membrane	Care and prevention of leukemia and cancer
Hydrolysis of blood proteic toxins (trysin, pronase)	Entrapment in polyuretane membrane	Removal of blood toxins in dialytic patients
Insulin secretion (Islets of Langerhans)	Polyamide UF membrane	Bioartificial pancreas
Hydrolysis of whey proteins (trypsin, chymotrypsin)	Polysulfone UF membrane	Production of peptides for medical use
Liver metabolites (hepatocites)	Polysulfone and polyamide MF membrane	Bioartificial liver

Table I.4 Membrane reactors in agro-food industry

Reaction	Membrane bioreactor	Purpose
Hydrolysis of lactose to glucose and β-galactose (β-galactosidase)	Axial-annular flow reactor	Delactosization of milk or whey for human consumption
Hydrolysis of raffinose (α-galactosidase and invertase)	Hollow fibre reactor with segregated enzyme	Production of monomeric sugars
Hydrolysis of high molecular mass protein in milk (trypsin and chimotrypsin)	Asymmetric hollow fibre with gelified enzyme	Production of baby food
Hydrolysis of starch to maltose (α-, β-amilase, pullulanase)	CSTR with UF membrane	Production of syrups 42 DE and HFCS
Fermentation of all fermentable sugars (yeast)	CSTR with UF membrane	Brewing industry
Anaerobic fermentation (*S. Cerevisiae*)	CSTR with UF membrane	Production of alcohol
Hydrolysis of pectines (pectinase)	CSTR with UF membrane	Production of bitterness and clarification of fruit juice and wine
Fermentation of *L. bulgaricus*	CSTR with UF membrane	Continuous fermentation for production of carboxylic acids
Removal of limonene and naringin (β-cyclodextrin)	CSTR with UF membrane	Production of bitterness and clarification of fruit juice
Hydrolysis of K-casein (endopeptidase)	CSTR with UF membrane	Milk coagulation for dairy products
Hydrolysis of collagen and muscle proteins (protease, papain)	CSTR with UF membrane	Tenderization of meat and particular beef
Conversion of glucose to gluconic acid (glucose oxidase/catalase)	Packed bed reactor	Prevention discoloration and off-flavor of egg product during storage
Hydrolysis of triglycerides to fatty acids and glycerol (lipase)	UF capillary membrane reactor	Production of food, cosmetics, emulsificant
Hydrolysis of cellulose to cellobiose and glucose (cellulase/β-glucosidase)	Asymmetric hollow fibre reactor	Production of ethanol and protein
Hydrolysis of malic acid to lactic acid (*Lactobacillus oenos*)	MF capillary membranes with entrapped cells	Improve test in white wine

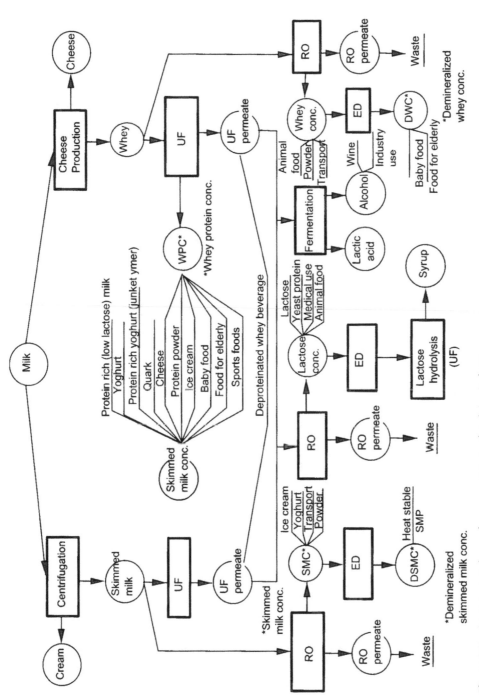

Figure I.2 An integrated membrane system in the dairy industry.

In this book we will analyze the various basic aspects and applications of catalytic membrane reactors in relation to their utilization in the production of bioactive species. Some general principles of membrane preparation and of the various available membrane separation operations will be also discussed.

References

DRIOLI, E., GIANFREDA, L., PALESCANDOLO, R. and SCARDI, V., 1975, in THOMAS, D. and KERVENER, J.P. (eds) *The Kinetic Behaviour of Enzymes Gelified on UF Membranes in Analysis and Control of Immobilized Enzyme Systems*, North Holland, American Elsevier, New York.

DRIOLI, E., NATOLI, M., KOTER, I. and TROTTA, F., 1995, An experimental study on a beta-cyclodextrin carbonate membrane reactor in PNPA hydrolysis. *Biotechnol. Bioeng.*, **46**, 415–420.

GERLACH, J.C., 1996, Development of a hybrid liver support system: a review, *Int. J. Artif. Organs*, **19**(11), 645–654.

HSIEH, H.P., 1996, Inorganic membranes for separation and reaction, *Membrane Science and Technology Series*, **3**, Elsevier, Amsterdam.

KIKUCHI, E., 1997, Hydrogen-permselective membrane reactors, *Cattech*, **67**, p. 74.

MESSING, R.A., 1975, *Immobilized Enzymes for Industrial Reactors*, MESSING, R.A. (ed.), Academic Press, New York.

ZAKS, A. and KLIBANOV, A.M., 1988, Enzymatic catalysis in non-aqueous solvents, *J. Biol. Chem.*, **263**, 3194–3201.

General Principles

General Principles

1

Membranes and Membrane Processes

The increasing and broad interest in membranes and membrane operations results largely from their basic properties, already observed and tested in biological systems. Membrane separations in general do not involve phase changes; they are athermal and work at room temperature; they are non-destructive and do not require the addition of other chemicals. Chemical engineers and physical chemists have been able to use these properties with intelligence in developing molecular separation operations and chemical reactors with innovative characteristics. Final users of membranes will use them as prepared and will not be involved in their preparation. However, knowledge of their composition and configuration is important for their correct use. Flat sheet membranes, hollow-fibre or tubular membranes can be prepared starting from the same materials. The different final configurations and their modular assembly are directly related to their overall performance in the membrane process plant.

'A membrane is a structure separating two phases and/or acting as an active or passive barrier to the transport of matter between the phases adjacent to it' (Koops, 1995). This definition can also be extended to the transport of energy as in photosynthetic membranes. A membrane is in general a solid or a liquid film, of small thickness if compared to its surface, which might ideally represent an interface between two phases. The most important characteristic of a membrane is its selective transport properties. This selectivity, combined with a significant permeability, results from the intrinsic chemical properties of the materials forming the membrane and its physical properties. Solubility and diffusivity of the penetrants in the dense polymeric phase, and the presence and size of micropores or of electrical charge, control the membrane transport phenomena.

Most commercial membranes are formed from organic polymers or from ceramics. Synthesized metal or glass membranes have also been produced and utilized in specific processes. The earliest synthetic membranes were based on derivatives of naturally occurring cellulosic materials, such as the asymmetric reverse osmosis membrane prepared by Loeb in the early 1960s (Loeb and Sourirajan, 1962; Loeb, 1993). New materials are now available, such as polyamide, polysulfone and polypropylene. Table 1.1 summarizes some of the most common polymeric materials used, particularly in pressure-driven membrane processes.

Table 1.1   Most common polymeric materials for membrane preparation

Material	Application					
	MF	UF	NF/RO	GS	PV	MD
Cellulose acetate	×	×	×	×	×	
Cellulose triacetate	×	×	×			
Blend CA/triacetate			×			
Cellulose esters	×					
Cellulose nitrate	×					
Blend CA/CN	×					
Poly(vinyl alcohol)	×					
Polyacrylonitrile		×			×	
Poly(vinyl chloride)	×					
PVC copolymer	×	×				
Acrylic copolymer	×					
Aromatic polyamide	×	×	×			
Aliphatic polyamide	×	×				
Polyimide	×	×	×	×		
Polysulfone	×	×				
Sulfonated Polysulfone		×	×	×		
Polyetheretherketone (PEEK)	×	×		×		
Polycarbonate	×					
Polyester	×					
Polypropylene	×				×	×
Polyethylene	×				×	×
Polytetrafluoroethylene (PTFE)	×	×			×	
Poly(vinylidene difluoride) (PVDF)	×	×			×	×
Collagen					×	
Chitosan					×	
Zeolytes				×	×	
Polyorganophosphazene				×	×	
Polydimethylsiloxane (PDMS)				×	×	

Dense homogeneous membranes do not present any detectable pore at the limits of electron microscopy (<5 nm) throughout their thickness.

Symmetric-microporous membranes are characterized by a clear pore structure with differing morphology depending on the preparation technique. Typical structures are presented in Figures 1.1 and 1.4, where a polypropylene membrane obtained by thermal phase inversion and a PTFE membrane obtained by mechanical stretching, respectively, are shown.

Asymmetric membranes, the most interesting for industrial applications in reverse osmosis, gas separations, ultrafiltration, and so on have a thin, dense layer in the top region over a more macroporous supporting layer. The asymmetric structure can be obtained in a simple film by phase inversion, or in composite systems by forming a thin, dense layer, for example by polycondensation *in situ* on macroporous membranes prepared from a different material. A removable top layer can also be realized dynamically (van Henven and Bloebaum, 1974).

Ion exchange membranes (electrically charged membranes), are characterized by a certain charge density related to fixed ionized groups in the polymeric chain. The charge

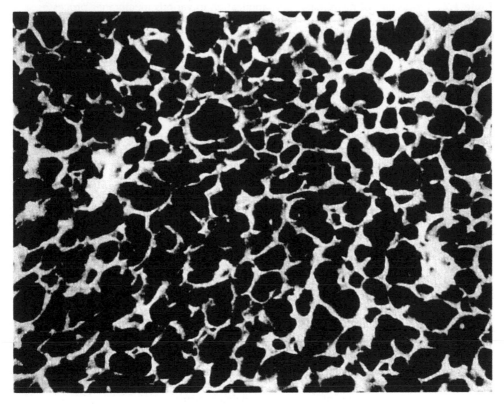

Figure 1.1 Polypropylene membrane obtained by thermal phase inversion.

density can be the origin of the membrane selectivity, as for electrodialysis membranes, or can contribute to it, as in ultrafiltration and nanofiltration.

Table 1.2 summarizes how the different types of membranes are produced and their application to various membrane processes. Membrane structure and preparation techniques will be described in the following section.

1.1 Membrane Structure

1.1.1 Dense Homogeneous Membranes

Dense membranes are produced by melt extrusion, solution techniques such as casting and careful evaporation of solvent (to avoid pore formation) and by direct polymerization and cross-linking. Such membranes can be visualized as a close assemblage of polymer chains in which the only free volume is that between the segments of these chains. The permeability of such a membrane will obviously depend on the ability of the permeant to enter the polymer matrix and then move through it, that is its solubility and diffusivity within the membrane.

An important consideration is the degree of mobility of the polymer chain segments since dynamic free-volume will encourage permeant transport. Thermal motion of chain segments will only occur in amorphous regions of the polymer, and permeation will be negligible in the denser, more ordered, crystalline regions. Polymers tend to be a mixture

Table 1.2 Production of membranes and application to various membranes processes

Membrane type	Method of preparation	Structure	Membrane process
Symmetric-microporous	• Stretching, hot and cold (PTFE)	Random network, pores 0.02–10 µm diameter	Microfiltration
	• Irradiation and track-etching (polycarbonate)	Parallel pores, 0.03–8 µm	Microfiltration
	• Casting and phase inversion (cellulose esters, polypropylene)	Random pores 0.1–1 µm	Microfiltration, dialysis
	• Molding and sintering (ceramics, PTFE)	Random pores 0.1–20 µm	Microfiltration
Asymmetric			
Single component	Casting and phase inversion (cellulosics, polyamides, polysulfone)	Dense or finely-porous skin grading of (macro) microporous substructure	Reverse osmosis, ultrafiltration, microfiltration, gas permeation
Composite	Film formation on microporous support	Dense skin on microporous sublayer	Reverse osmosis, gas permeation
Dynamic	Deposition of fine precoat on microporous barrier	Thin (removable) layer on microporous membrane	Reverse osmosis, ultrafiltration
Electrically charged	Sulfonation and amination of homogeneous dense membrane	Fixed charge groups in polymer matrix	Electrodialysis

of amorphous and crystalline regions, and as temperature is lowered the crystallinity increases. Temperature changes also influence the motion of chain segments in the amorphous region. Below the glass transition temperature the motion ceases, so that permeation is greatly reduced in 'glassy' polymers.

Dense membranes have the disadvantage of low flux unless they can be made extremely thin. For this reason, dense membrane properties are incorporated into the top 'skin' layers of asymmetric membranes. It should be noted that the earlier definition of no detectable pores allows for the presence of ultrafine pores (<5 nm), and there is some evidence of these in the 'dense' skins of asymmetric membranes.

1.1.2 *Symmetric-Microporous Membranes*

Microporous membranes have readily detectable pores or voidage. As indicated in Table 1.2, these membranes are primarily used for microfiltration, including crossflow microfiltration. They are also used as the substrate layer in composite membranes, as a support matrix for liquid membranes, in membrane distillation, and so on. Several methods have been developed to produce microporous membranes, and each method yields a different microstructure as shown in the cross-section obtained by electron microscopy. Figure 1.1 shows the porous structure of a polypropylene membrane obtained by thermal phase inversion.

Figure 1.2 Nucleopore microporous membrane with cylindrical pores obtained by 'track-etching'.

The simplest form of microporous membrane is a dense polymer film with cylindrical pores or capillaries, as shown in Figure 1.2, produced by Nucleopore Corp. using the 'track-etching' procedure.

Figure 1.3 depicts this method of preparation, which involves irradiation of a polycarbonate film (~10 µm thick) followed by chemical etching of the tracks formed by radiation damage. Irradiation time controls pore density and etching time controls pore size, which can range from 30 to 80 nm. The relatively low void volumes of the track-etched membranes give them low filtration rates, and for this reason they are unsuitable for many general applications. All other microporous membranes have a more open and random structure. Figure 1.4 shows a membrane produced from polytetrafluoroethylene (PTFE). Sintering of a PTFE powder produces a granular structure with interconnected tortuous pores of varying diameter, the dimensions being determined by the particle size of the initial powder. Similar membranes can be produced from metals and ceramics (Norton Co. produced such a membrane in 1985), although modern ceramic membranes tend to have an asymmetric structure. Sintering has the disadvantages that the void volume is limited to about 40% and fine pores are difficult to form.

Microporous membranes can be prepared by controlled stretching and tearing of dense polymer films. Their microstructure has high voidage and random tortuous paths of characteristic size. Examples of this type of membrane are the Gore-tex (W.L. Gore) PTFE membrane, the Celgard (Celanese) and polypropylene membrane.

Phase inversion is the most important method for production of microporous membranes and will be discussed in detail in the following section.

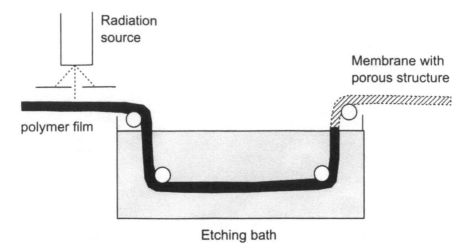

Figure 1.3 'Track-etching' procedure for obtaining microporous membranes.

Figure 1.4 Polytetrafluoroethylene membrane.

1.1.3 Asymmetric Membranes

A major breakthrough in the development of membranes was the preparation by Loeb and Sourirajan (Loeb, 1993) of asymmetric membranes (the term 'anisotropic' is used interchangeably). The term asymmetric refers to membranes formed by a porous spongy wall supporting a very thin dense layer. The thin skin layer, approximately 0.5 μm

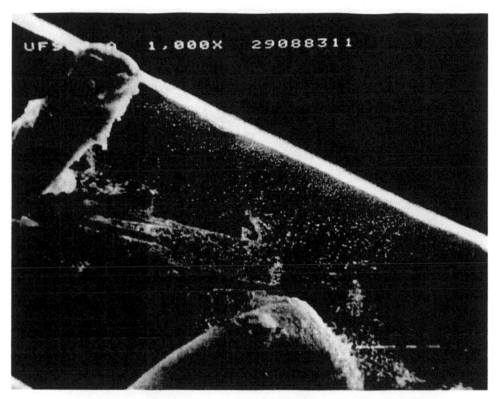

Figure 1.5 Asymmetric membrane with a thin skin above a microporous structure.

thick, with pore sizes in the range 0.001–0.2 μm (1–200 nm), determines the separation properties of the membrane with little hydraulic resistance to mass transport. This skin layer is fully surrounded by a spongy structure, approximately 75 μm thick, with pore sizes of 5–10 μm and 80–90% porosity, which has a very high hydraulic permeability. This porous spongy wall acts as a mechanical support for the semipermeable thin layer.

Figure 1.5 depicts an asymmetric membrane with a thin (0.1–1.0 μm) 'skin' above an open microporous structure about 100 μm thick. For reverse osmosis membranes, and some forms of gas permeation membrane, the skin will have no detectable pores, being either dense polymer or finely porous. For ultrafiltration the 'skin' will be porous but will contain pores that are significantly smaller than the voids in the substructure. The important point is that the 'skin' layer provides the separation capability of the membrane and its thickness minimizes the resistance to flow. Asymmetric membranes are prepared either by the phase-inversion process or by forming a multilayer composite. They can also be prepared in a hollow-fibre configuration (Figure 1.6).

1.1.4 Composite Membranes

The advantages of asymmetric membranes can be achieved by producing a thin-film composite (TFC) from a microporous substrate overlain with an ultra-thin active layer.

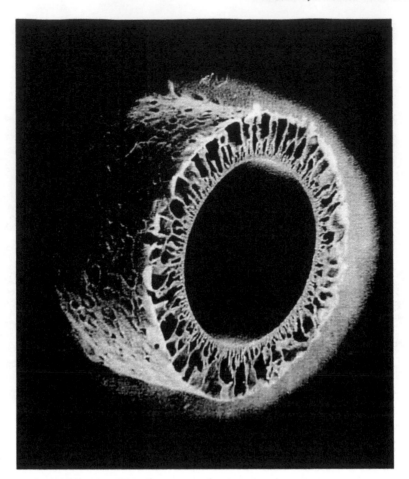

Figure 1.6 Hollow-fibre configuration.

This approach has additional advantages since the support and film can be 'tailored' to the needs of the process, and because the problem of gradual skin growth and densification (leading to flux decline) that occurs with conventional asymmetric membranes does not take place (Rozelle et al., 1975).

The substrate is basically an ultrafitration membrane made from chemically resistant polysulfone. A thin coating of polyethylenimine (PEI) is applied, which is heat cured to make it water insoluble and cross-linked with toluene 2,4-diisocyanate (TDI) to make it more dense and mechanically stable. An overview of the preparation of composite reverse osmosis membranes has been given by Cadotte and Peterson (1981).

A rather different concept is behind the composite membrane developed by Monsanto for gas separations (Henis and Tripodi, 1981). This membrane has a polysulfone substrate coated with a thin film of silicon rubber that plugs and penetrates into the pores and micropores of the polysulfone. In this case polysulfone provides the selectivity and the silicon rubber acts to prevent permeant by-passing.

A *dynamic membrane* is a composite of substrate (microporous) membrane and deposited film. The term dynamic refers to the fact that the film is deposited *in situ* and can be removed and replaced as required.

1.1.5 *Methodology for Membrane Preparation by Phase Inversion*

This technique is of great importance in the production of commercial membranes and has been explained in detail by several workers including Kesting (1971) and Pusch and Walch (1982). The phase-inversion process has been described as a sol to gel transition. Thus the dope is a colloidal dispersion, or sol, which is transformed by removal of solvent into a colloidal network, or gel, which ultimately forms a solid matrix, the membrane. Figure 1.7 depicts the membrane fabrication process. The dope formulation will contain polymer and solvent, and may include a swelling agent and/or a non-solvent as well as ingredients, such as lithium or magnesium salts, that influence the kinetics of the phase separation process. It is usually important that the solvent is the most volatile component.

After the dope has matured for a period of time, it is cast as a film, either as a flat sheet on a smooth surface or inside a porous tube, or extruded through a spinnerette as a hollow fibre. Solvent is allowed to evaporate for a brief or extended period depending on the desired effect. Sol to gel transition may occur at this point. The film is then quenched in non-solvent, typically water, which completes the phase inversion. Washing is required to remove residual solvent and additives and, in some cases, post-treatment by heat annealing is important.

An example of a recipe for an asymmetric reverse osmosis membrane made from cellulose acetate is given in Table 1.3 (Brock, 1983). In the process described, the acetone is a solvent for cellulose acetate and the water is the non-solvent. During the evaporation period the more volatile acetone becomes depleted at the dope–air interface and the polymer content increases, forming a thin skin layer above the fluid interior. The phase separation is completed during quenching as the solvent and non-solvent diffuse countercurrently through the skin and the polymer slowly forms a solid sponge-like substructure.

Post-treatment by heat annealing causes polymer molecules and polymer clusters in the skin to realign, decreasing the effective void space, be it intermolecular or fine or coarse pores. As a result the membrane shows increased rejection and decreased flux. This annealing effect is marked for cellulosic materials but not significant for others, such as aromatic polyamides (Pusch and Walch, 1982).

The microporous substructures of asymmetric membranes may comprise sponge-like cells (Figure 1.8) or elongated fingers (Figure 1.9). Sponge-like structures have better mechanical properties and are desirable for reverse osmosis membranes, which must withstand considerable pressure. Fingers are commonly found in the substructures of ultrafiltration membranes.

The recipe for a phase-inversion membrane in Table 1.3 was obtained by empirical means, guided by statistically designed experiments. Because of the large number of variables, in the preparation of phase-inversion membranes it is not possible with any certainty to predict the membrane structure from a new formulation. However, some qualitative guidelines are summarized in Table 1.4.

Another phase-inversion technique which produces truly symmetric microporous membranes is thermal phase inversion, an example of which is the Accurel process (Schneider, 1981) for polypropylene (PP) developed by Enka. In this process PP is mixed with N,N-bis(2-hydroxyethyl)hexadecylamine to form a binary system which is miscible at elevated temperature (150 °C) but immiscible at lower temperature (50 °C). Phase separation is obtained by extruding the note dope, as films, tubes or fibres into a cooled bath of the hexadecylamine. This rapid quenching produces a very symmetric microstructure, with a relatively narrow distribution of effective pore size. A necessary final step involves washing the membranes to remove solvent.

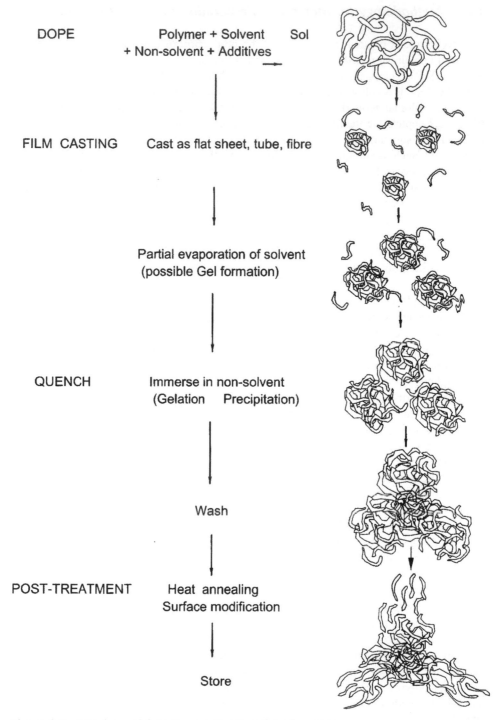

DOPE   Polymer + Solvent   Sol
                  + Non-solvent + Additives

FILM CASTING   Cast as flat sheet, tube, fibre

                  Partial evaporation of solvent
                  (possible Gel formation)

QUENCH   Immerse in non-solvent
                  (Gelation   Precipitation)

                  Wash

POST-TREATMENT   Heat annealing
                  Surface modification

                  Store

Figure 1.7   Membrane fabrication process by phase inversion.

Table 1.3 Recipe for asymmetric membrane

	Component	Proportion (wt%)
Dope preparation	Cellulose nitrate (11% nitrogen)	5.5 (dried)
add	Ethyl alcohol (absolute)	9.0 (24 h stand)
add	Ether (anhydrous)	27.0 (stir to dissolve)
add	Acetone (anhydrous)	41.5 (2 h stir)
add	Amyl alcohol (anhydrous)	17.0 (2 h stand)
Casting	Form film 200–700 μm on glass plate	
Precipitate	Hold for 75 min, 22–24 °C, relative humidity 55–60% in water bath at 22–24 °C	

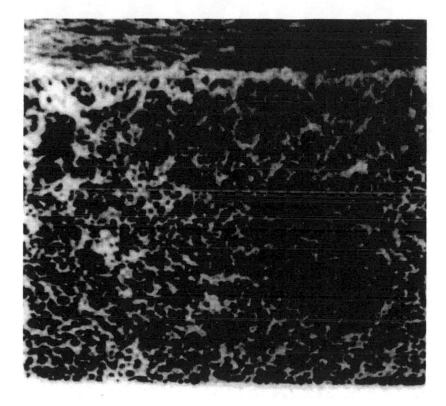

Figure 1.8 Sponge-like microporous substructure of an asymmetric membrane.

1.2 Membrane Materials

Early membrane developments used derivatives of cellulose. Membranes are made from cellulose esters, which are obtained by reaction of cellulose with the appropriate acid (Figure 1.10).

Cellulose acetate (CA) with between 3 and 2.5 acetyl units, is still used for many types of reverse osmosis membrane. The advantages of CA are its excellent film-forming

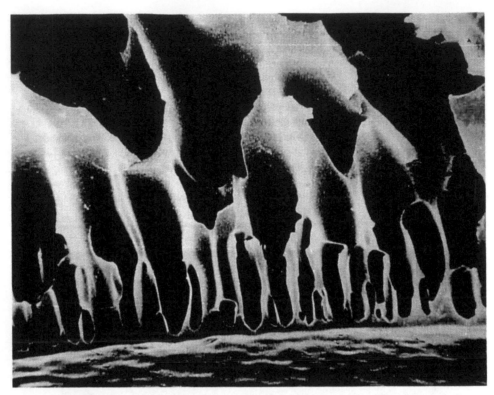

Figure 1.9 Elongated-finger microporous substructure of an asymmetric membrane.

Table 1.4 Summary of qualitative guideline for phase-inversion process

Method	Membrane structure
A. **Minimization of concentration gradients as dope approaches phase inversion**	Symmetric-microporous
(A.1) Gelation with slow input of non-solvent e.g. Non-solvent absorbed from vapour phase	
(A.2) Weak solvent and/or weak precipitating agent	
(A.3) Rapid transition of solvent to a non-solvent e.g. Thermogelation	
B. **Induce steep concentration gradient in the surface of the dope**	Asymmetric
e.g. Partial evaporation of solvent before precipitation	
(B.1) Induce rapid phase inversion (e.g. decreasing solvent/swelling agent ratio, or using 'powerful' non-solvent)	Increased asymmetry, more coarsely porous, with finger formation
(B.2) Induce slow phase inversion (e.g. increase polymer concentration or viscosity)	Decrease in asymmetry, more finely porous, with sponge structure

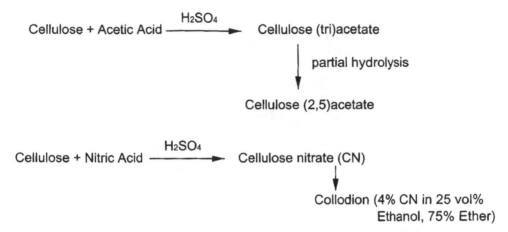

Figure 1.10 Membrane manufacture from cellulose esters: reaction of cellulose.

properties, with dense skin and spongy substructure, its good rejection of salts and its water flux. Cellulose esters or mixed esters are also commonly used for microfilters. Dialysis membranes are often formed from regenerated cellulose that is obtained from a cellulose acetate membrane converted back to cellulose by treatment with alkali.

The major disadvantage of cellulose acetate is its susceptibility to hydrolysis, which limits usage to a rather narrow pH range (5–7). A large number of more robust alternative polymers are now used for membrane fabrication. For reverse osmosis the use of aromatic polyamides in hollow fine fibre form has proved very successful, and for ultrafiltration many systems now use polysulfone.

Inorganic membranes became commercially available in the late 1980s for practical applications of membrane reactor technology; membranes should have high permeability and good separation selectivity. They must also be stable at reaction temperatures and able to withstand a significant pressure drop.

Inorganic membranes, like organic ones, can be divided first into nonporous (dense) and porous, and secondly into supported (asymmetric) and unsupported (symmetric). Nonporous metal membranes such as palladium alloys, which absorb hydrogen atoms and then transport them by diffusion, were the first type used to combine heterogeneous catalytic reaction and separation (Basile et al., 1996). Virtually complete reaction conversion is possible with these membranes because they allow only one component to permeate. Nonporous oxide membranes have also been used. The porous membranes are glasses with small pores, composite ceramics, and zeolites.

Selectivity generally grows as pores become smaller, and it is virtually infinite for dense membranes; generally, as selectivity or membrane thickness rises, permeability decreases. Improvements in thin-film deposition techniques (chemical vapour deposition, sputtering, co-condensation, etc.) made it possible to produce asymmetric inorganic membranes in which a thin layer gives selectivity without lowering permeability too much, while a support, generally a macroporous sintered ceramic, assures mechanical strength. Below a critical thickness of the selective layer, defects such as a cracks and pinholes become technically unavoidable and cause an unacceptable decrease in selectivity. High cost, fabrication durability and poisoning (by carbon and sulfur compounds) are the most important problems these membranes present. The metal membranes used are palladium and silver.

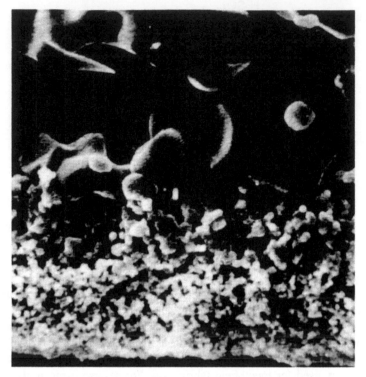

Figure 1.11   Asymmetric ceramic membrane.

Nonporous oxide membranes have seen only limited use, though some oxides exhibit high separation selectivities for H_2 or O_2. For example, silica is permeable to H_2 which moves through the openings in the silica network. Solid electrolyte membranes, such as calcium-stabilized zirconia, are selective for O_2 but not to other gases, and silver has been used to selectively permeate O_2 (Basile et al., 1996).

Uniform, microporous Vycor glass membranes can be prepared with pores as small as 4 nm. These glass membranes may be limited in their applications, particularly because they are brittle. Moreover, when microporous glass is heated above 575 K for long periods, or to much higher temperatures for shorter periods, it loses its microstructure. Some glasses are stable above 1075 K, however. The development of composite ceramic membranes, with pore diameters as small as 2.5 nm, may also limit the use of microporous glass membranes because higher fluxes can be obtained through microporous ceramic membranes. The ceramics are also stable to high temperatures (alumina membrane can be used to 1075 K without degradation of the pore structure) and mechanically stable and can withstand pressure drops of 1.5 MPa and they are resistant to corrosive chemicals. Ceramic membranes have controlled, stable and narrow pore-size distribution. Catalytic materials that are deposited by impregnation can also be dispersed on ceramics and thus they can have high catalyst surface/volume ratios (Basile et al., 1996). Most ceramic membranes consist of a layered or graded structure with a thin (few micrometres) permeselective layer deposited onto a thicker (several micrometres) macroporous layer that is essentially a high-temperature support (Figure 1.11).

To prepare nanometre size particles required for making membranes with pore diameters of a few nanometres, which correspond to the large majority of the present commercial products, the sol–gel process is often employed.

Table 1.5 Comparison of different types of modules applied to liquid processing

Characteristic	Module concept				
	Flat-plate	Spiral wound	Shell and tube	Hollow fibre	Rotary disk
Packing density (m^2/m^3)	200–400	300–900	150–300	9000–30 000	—
Flux ($l\ m^{-2}\ h^{-1}$)	10–50	10–50	10–50	0.5–5	30–60
Flux density ($l\ m^{-3}\ h^{-1}$)	2000–20 000	3000–45 000	1500–15 000	4500–150 000	650–1200
Channel diameter or height (mm)	5	1.5	13	0.1	—
Membrane replacement	Sheets	Module cartridge	As tubes	As module	Sheets
Replacement labour	High	Medium	High	Medium	—
Feed side pressure loss	Medium	Medium	High	Low	Low
Concentration polarization	Medium	Medium	High	Low	Low
Suspended solids capability	Medium	Poor	Good	Poor	High
Cleaning frequency	Medium	Medium	Low	High	Very Low

During the 1990s, Tech-Sep initiated an ambitious R&D project to design inorganic nanofiltration membranes. These membranes were first developed in collaboration with the nuclear industry (Commissariat à l'Energie Atomique and its subsidiary SFEC, France). The membrane is a pure inorganic ZrO_2 layer obtained by the sol–gel technology and deposited on a ceramic support Kerasep™. New possibilities have been opened by nanofiltration ceramic membranes that have now been available on the market for several years: they can be used in a very broad range of operating conditions (pH from 0 to 14, severe oxidizing or reducing conditions, thermal resistance from 0 to 350 °C, high pressure resistance, inertness towards radiation, etc.), which means that new membrane applications can be examined.

Pure zeolites are crystalline; this allows a full regeneration in case of fouling but makes them brittle and easily affected by cracks. An interesting approach to the synthesis of zeolite membranes is based on letting crystals grow close to one another on a mesoporous support (Geus and van Bekkum, 1989; Geus et al., 1991; Jia et al., 1994). Some methods of preparation of zeolite membranes are reported in recent reviews (Liu et al., 1996; Boom et al., 1998).

1.3 Modules for Membranes

A module is a unit in which the membrane is housed. In the module the feed is divided into the permeate — the part that passes through the membrane — and the retentate.

The most common membranes are flat or tubular. 'Plate and frame' and 'spiral-wound' modules use flat membranes, while tubular membranes and hollow fibres are assembled in different numbers (from a few 3/8-inch tubes to thousands of 40 µm fibres) in a multitube heat exchanger configuration. The choice among different modules is made on the basis of factors such as cleanability and ease of replacement of the membrane.

A comparison of the different module concepts as applied to liquid processing is presented in Table 1.5.

The main purposes of the module are:

- to provide mechanical support for the membrane;
- to permit effective control of boundary layers and of concentration polarization phenomena, particularly in liquid phase processes;

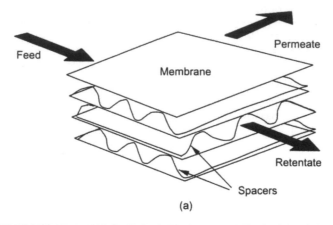

(a)

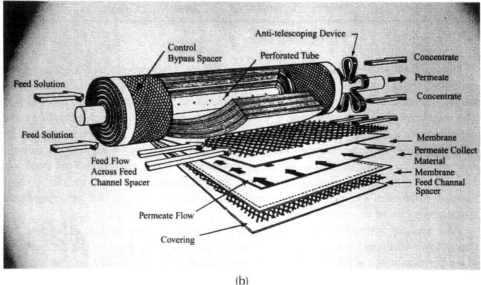

(b)

Figure 1.12 (a) Plate-and-frame module. (b) Spiral-wound membrane module.

- to give a high packing density of membrane area per unit volume;
- to permit easy cleaning of the membranes;
- to facilitate their maintenance and replacement.

The basic properties for each concept are as follow.

Plate and frame. This system is based on the plate and frame concept used in conventional filtration. Figure 1.12a depicts one form of flat plate system, which is simply a stack of membranes, on porous supports, and spacers arranged so that feed-fluid flows on one side of the membrane and on the other side there is either permeate flow or the flow of a receiving fluid. For example, in electrodialysis, which commonly uses the flat plate arrangement, the feed stream is depleted of ions and the receiving stream becomes concentrated. The flat plate concept usually employs thin channels so that laminar flow occurs. In this system, it is possible to have 100–400 m^2 of membrane area per m^3 module volume. Membrane replacement in the flat plate module is simply a matter of replacing

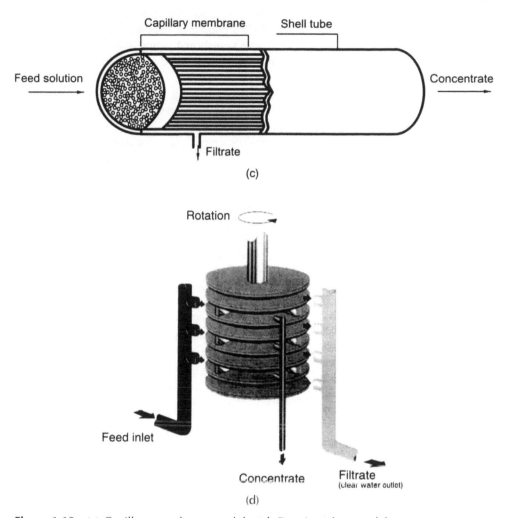

Figure 1.12 (c) Capillary membrane module. (d) Rotating plate module.

membrane sheets, but the economic advantage of this has to be weighed against the high labour costs.

Spiral-wound. An extension of the flat plate system is to wind the membrane, the porous support, and the spacer in a spiral configuration as shown in Figure 1.12b. The membrane element is therefore cylindrical and fits tightly into a cylindrical housing. Feed flows parallel to the axis of the cylinder and permeate flows through the membrane into the porous support and then around the spiral into the permeate outlet pipe that runs along the axis of the cylinder. It can be seen that the flow channel for the feed is a curved slit. The spiral-wound module has a cartridge shape that allows $300-1000 \ m^2$ of membrane area per m^3 module volume.

Hollow fibres. 'Hollow fibre' or capillary membrane modules (Figure 1.12c) are essentially similar to multitube heat-exchangers. They allow around $30\ 000 \ m^2$ of membrane area per m^3 module volume but require clean feed to reduce the possibility of fouling the fibres. For gas feed, a pretreatment guarantees a good performance of the system but involves additional cost.

Spiral modules have advantages where pressure drop considerations are very important and when countercurrent flow is not needed to maximize separation efficiency. Higher pressure applications involving costly pressure vessels and piping will tend to favour the fibre format, which reduces this expensive component of the system by a factor of ten in some cases.

Rotating plate. In recent years, the interest in treating highly viscous systems has increased attention on rotating plate modules and spiral-wound modules with special spacers (Frenander and Jönsson, 1996) (Figure 1.12d).

The engineering of membrane systems makes a significant contribution to maximizing performance and cost. Generally, it is necessary to use more than one module to reach the required purity and yield. These modules have to be connected; normally, it is possible to dispose them in series or in parallel.

Parallel connection allows treatment at higher flow rates and, if all modules are the same, to divide the flow equally between each module (each module treats a flow rate equal to the effective rate divided by the number of modules in the parallel array). In this way, it is possible to obtain higher productivity. The net performance is the sum of the performance of the individual modules.

In series connection, each module works with a flow rate different from the previous one. This system increases product purity. Note that now the modules are in the same vessel, so there are lower costs and pressure drops. With modules connected in series there is a variety of possibilities for manipulating the various streams (feed, permeate and retentate) from each of the modules to optimize the effectiveness of the modules and hence the performance of the system.

Normally, a single stage without recycle is not as capital-intensive as a cascade with more stages owing to a lower membrane area requirement, but it gives lower purities and yields.

An economic analysis allows one to choose the best configuration for each implementation.

Figures 1.13a and 1.13b show, respectively, a module operating in a single permeation stage without recycle and two permeation stages in cascade with recycle. Hitachi, Japan, developed rotary multi-disk modules which adopt an innovative self-cleaning system featuring flat-sheet membrances attached to rotary disks positioned on different shafts. The rotating disks generate turbulence that agitates the membrane surface, producing a self-cleaning effect. As a result, it provides stable treatment over long periods of operation and reduces power consumption compared with the membrane processes which recirculate feed at high speed. The major application of these modules are waste water treatment, separation of solids and liquids in various types of industrial wastewater, and concentration application.

Rotary multi-disk modules allow 25–100 m² of membrane surface area for between two to four shafts and between 30 to 130 disk membranes. The dimensions (m) of modules are between $2 \times 2.7 \times 1.6$ (l·w·h) and $2.5 \times 4 \times 1.6$.

1.4 Membrane Operations

1.4.1 Crossflow Microfiltration

Microfiltration is a pressure-driven membrane operation in which particles are separated in the solvated size range of 50 to 10 000 nm from a fluid mixture. The term crossflow refers to a mode of operation where the feed flows parallel to the membrane surface and

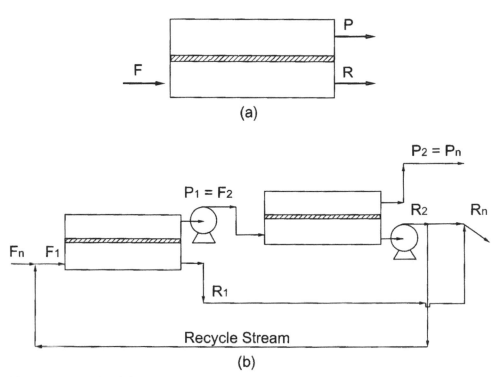

Figure 1.13 (a) Module operating in a single permeation stage without recycle. (F = feed stream, P = permeate stream, R = retentate stream). (b) Two permeation stages in cascade with recycle. (F1 = feed to first stage; F2 = feed to second stage; Fn = net feed; R1 = retentate stream from first stage; R2 = retentate stream from second stage; Rn = net retentate stream; P1 = permeate stream from first stage; P2 = permeate stream from second stage; Pn = net permeate stream).

the downstream fluid moves away from the membrane in a direction normal to the membrane surface. In microfiltration the flow of fluid through the filter medium is proportional to the applied pressure difference (ΔP) across the membrane. The rate of flow is dictated by the resistance of the membrane to the flow of fluid. The filter medium characteristics are defined in terms of its resistance or permeability. The flow rate or flux is thus also dependent on the viscosity, μ, of the fluid being filtered.

Filters for membrane microfiltration are typically made from thin polymer films with 'uniform' pore size and a high pore density of approximately 80%. The principal method of particle retention is 'sieving', although the separation is influenced by interactions between the membrane surface and the solution. The high pore density of the filters generally means that hydrodynamic resistance is relatively low and hence high flow rates or membrane flux rates (m^3 of permeate per m^2 of membrane area per hour, m/h) result at modest operating differential pressures up to 2×10^5 Pa.

The irregularity of the pores of most membranes and the often irregular shape of the particles being filtered mean there is no sharp cut-off size during filtration. With symmetric membranes some degree of in-depth separation occurs as particles move through the tortuous flow path. To counteract this effect, asymmetric membranes, which have surface pore sizes much less than those in the bulk of the membrane, can be used. These entrap the particles almost exclusively at the surface (the membrane skin) while still offering low hydrodynamic resistance. This technique has also enabled inorganic membranes to be used in several applications.

Microfiltration has traditionally been applied in a dead-end mode of operation. In this, feed flow is perpendicular to the membrane surface and retained (filtered) particles accumulate on the surface, forming a layer of retained solid or a filter cake. The thickness of this cake increases with time and the permeation rate decreases correspondingly as the resistance of the cake increases. Eventually the membrane filter reaches an impractical (uneconomic) flow filtration rate and is either cleaned or replaced.

1.4.2   Ultrafiltration (UF)

Ultrafiltration is a pressure-driven membrane operation in which the membrane fractionates components of a liquid predominantly according to their size and shape. The membranes used in ultrafiltration are characterized by pore diameters in the range 5–50 nm. Hence, ultrafiltration is a process of separating extremely small particles and dissolved molecules from fluids. The primary basis for separation is molecular size, although secondary factors such as molecular shape and charge can play a role. Materials ranging in molecular mass from 10^3 to 10^6 Da are retained by ultrafilter membranes, while salts and water will pass through. Colloidal and particulate matter can also be retained.

Ultrafiltration membranes can be used to purify, and collect, both the fluid material passing through the filter and material retained by the filter. Species smaller than the pore size rating pass through the filter and thus can be depyrogenated, clarified and separated from high molecular mass contaminants. Materials larger than the pore size rating are retained by the filter and can be concentrated or separated from low molecular mass contaminants.

Ultrafiltration membranes are operated in a tangential flow mode: feed material sweeps tangentially across the upstream surface of the membrane as filtration occurs, thereby maximizing flux rates and filter life. These systems offer the advantage of long life because ultrafilter membranes can be regenerated repeatedly with strong cleaning agents.

Osmotic effects in UF membranes are small and the applied pressure, of the order of 1×10^5 to 7×10^5 Pa, is primarily to overcome the viscous resistance of liquid permeation through the porous network of the membrane. Commercial UF membranes are asymmetric, with a thin skin, some 0.1–1 μm thick, of fine porous texture exposed to the feed side. This skin is supported on a highly porous layer some 50–250 μm thick, which combines to give the unique characteristic of high permeability and selectivity. Although most UF membranes are polymeric, inorganic ceramic membranes are now breaking into the market place. Typical membrane materials are polysulfone, polyethersulfone, polyacrylonitrile, polyimide, cellulose acetate, aliphatic polyamides and ceramics, for example zirconium and aluminum oxides.

The main hydrodynamic resistance of the membrane is in the membrane top layer, the supporting porous sublayer (or sometimes macrovoid layer) offers minimal hydraulic resistance. The value of permeability constant for UF membranes is much smaller than that for microfiltration membranes. The value of K permeability constant depends upon many structural properties. For pure water (or other liquids) there is a linear correspondence between flux and transmembrane pressure. With real solutions there is a tendency for flux to reach an asymptotic value with increasing pressure. This is a result of several factors, including concentration polarization, gelation, fouling and osmotic effects.

Examples of the main applications of UF are listed in Table 1.6.

Table 1.6 Basic properties of the most common membrane operation mode

Process	Concept	Driving force	Species passed	Species retained
Microfiltration (MF)	Feed → Retentate / Solvent — Microporous membranes	Pressure difference 100–500 kPa	Solvent (water) and dissolved solutes	Suspended solids, fine particulates, some colloids
Ultrafiltration (UF)	Feed → Retentate / Solvent — UF membranes	Pressure difference 100–800 kPa	Solvent (water) and low molecular mass solutes (<1000 Da)	Macrosolutes and colloids
Nanofiltration (NF)	Feed → Retentate / Solvent — NF membranes	Pressure difference 0.3–3 MPa	Solvent (water), low molecular mass solutes, monovalent ions	High molecular mass compounds (100–1000 Da), multivalent ions
Reverse osmosis (RO)	Feed → Retentate / Solvent — RO membranes	Pressure difference 1–10 MPa	Solvent (water)	Virtually all dissolved and suspended solids

Table 1.6 (cont'd)

Process	Concept	Driving force	Species passed	Species retained
Electrodialysis (ED)		Electric potential difference 1–2 V/cell pair	Solutes (ions), small quantity of solvent	Non-ionic and macromolecular species
Dialysis (D)		Concentration difference	Solute (ions and low molecular mass organics), small quantity of solvent	Dissolved and suspended solids with molecular mass > 1000 Da
Gas permeation (GS)		Pressure difference 0.1–10 MPA	Highly permeable gases and vapours	Less permeable gases and vapours
Pervaporation (PV)		Chemical potential or concentration difference	Highly permeable solute or solvents	Less permeable solute or solvents

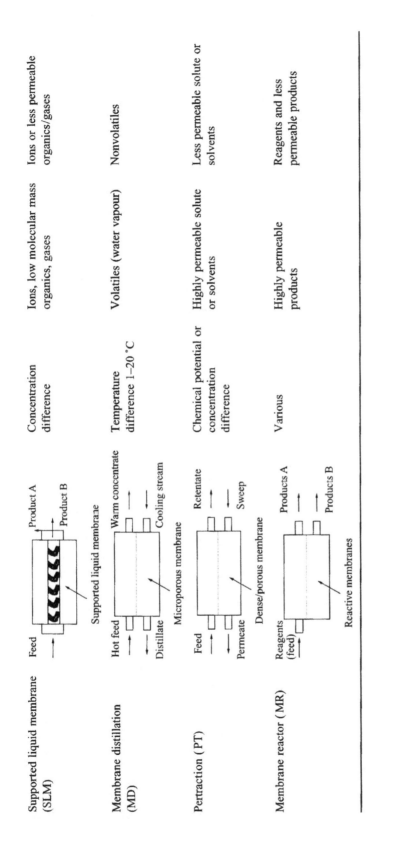

Supported liquid membrane (SLM)		Concentration difference	Ions, low molecular mass organics, gases	Ions or less permeable organics/gases
Membrane distillation (MD)		Temperature difference 1–20 °C	Volatiles (water vapour)	Nonvolatiles
Pertraction (PT)		Chemical potential or concentration difference	Highly permeable solute or solvents	Less permeable solute or solvents
Membrane reactor (MR)		Various	Highly permeable products	Reagents and less permeable products

1.4.3   Nanofiltration (NF)

Nanofiltration is a pressure-driven membrane operation in which the membrane fractionates components of a fluid predominantly according to their size and charge. It is mainly used for the separation of multivalent ions and uncharged organic molecules that have molecular masses between 100 and 1000 Da. The membranes used in nanofiltration are characterized by a charged surface and equivalent pore diameters in the range 1–3 nm.

It has only recently achieved success as a result of development of thin-film non-cellulosic membranes. Membranes can be formed by interfacial polymerization on a porous substrate of polysulfone or polyethersulfone. Generally, this opens up the possibilities for improvements in process efficiency and the production of new products, particularly in the food and biotechnology industries. Nanofiltration systems typically operate at lower pressures than RO (e.g. 5×10^5 Pa) but yield higher flow rates of water, albeit of a different quality from RO.

Nanofiltration is used when high sodium rejection, typical of RO, is not needed but where other salts such as Mg^{2+} and Ca^{2+} (i.e. divalent ions) are to be removed. The molecular mass cut-off of the NF membrane is around 200 Da. Typical rejections, defined as (Cf-Cp)/Cf, where Cf is the feed concentration and Cp is the product concentration, are (5×10^5 Pa, 2000 ppm solute) 60% for sodium chloride, 80% for calcium bicarbonate and 98% for magnesium sulfate, glucose and sucrose.

1.4.4   Reverse Osmosis (RO)

If two aqueous solutions of different concentrations are separated by a semipermeable membrane, there is a flux of water through this membrane from the less concentrated solution to the more concentrated one. The flux stops when the difference of level between the two solutions becomes equivalent to the difference between their osmotic pressures. If we impose on the more concentrated solution an external pressure higher than the difference between the two osmotic pressures, we will have a flux of water from the more concentrated solution to the less concentrated one. This phenomenon, termed reverse osmosis, can be used for the desalination of water.

Reverse osmosis membranes can essentially separate all solute species, both inorganic and organic, from solution. The mechanism of separation of species is based on processes relating to their size and shape, their ionic charge and their interactions with the membrane itself. This mechanism can be visualized as thermodynamically controlled partitioning, analagous to solvent extraction. The operating principle, referred to as the solution–diffusion model, is that a surface layer of the membrane is a relaxed region of amorphous polymer in which solvent and solute dissolve and diffuse. To overcome the molecular friction between permeates and membrane polymer during diffusion, large operating pressures are required in the range of 3×10^6 to 1×10^7 Pa.

The membranes used for reverse osmosis are either asymmetric or composite with typically a dense top layer <1 μm thick supported by a 50–150 μm-thick porous sublayer. The top layer imparts the intrinsic separation characteristics and the thinness of this layer ensures high flux rates. Typical membranes are made by phase inversion of cellulose triacetate and aromatic polyamides and from other polymers using interfacial polymerization (composite membranes).

The major application of RO membranes is in the processing of aqueous solutions containing inorganic solutes. (See Table 1.6). The particle size range for applications of RO is approximately 0.1–1 nm, and with solutes of molar masses greater than 300 Da

Table 1.7 Types of separation for which RO is used in food processing

Type of separation	Example
Water treatment	Pretreatment of boiler water
	Water softening
	Recycle of hot process fluid
Product and chemical recovery	Recovery of sugars and acids from rinse waters from fruit cocktail dicer
	Recovery of peach by-products
	Recovery of caustic for peeling operations
	Regeneration of cleaning solutions and sanitizers
	Recovery of sweet-potato stillage
Concentration/denaturing	Juices (e.g. tomato, orange, apple, grape)
	Water from processing of fish
	Concentration of milk or whey for cheese production or for transport
	Maple syrup
Fractionation	Fruit juice clarification
	Recovery of flavours, fragrances, pectins and proteins
	Removal of limonine from orange juice
	Separation of sugars from proteins in tomato serum
	Removal of alcohol from wine
	Removal of citric acid from pasta-blancher water

complete separation is achieved. RO offers advantages over competing technologies, including low energy requirements, low processing temperatures (and minimization of thermal damage to chemicals during processing), continuous rather than batch operation, modular construction and simple system designs.

Food processing applications such as dewatering and the concentration of foodstuffs were among the first uses of RO technology. Food processing applications were a logical use of RO because, where practical, the technology offers many potential advantages over conventional technologies used by the food processing industry. Table 1.7 lists examples of the main types of RO applications in food processing and examples for each type of separation.

Reverse osmosis technology is now being used in meat and seafood applications, in fruit and vegetable processing, and in the beer and wine industries.

1.4.5 Pervaporation (PV)

Pervaporation is a membrane operation driven by a chemical potential difference across the membrane in which the feed and the retentate streams are both liquid phases, while the permeate is a vapour.

In the pervaporation operation, the liquid feed mixture is maintained in contact with one side of a membrane and the permeate is continuously removed from the other side in vapour form by a vacuum pump. A selective dense membrane can modify the composition of the vapour–liquid equilibrium that can be established among the components of

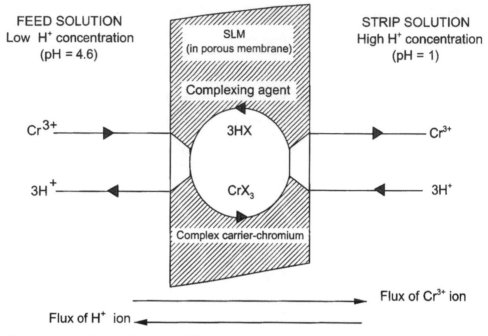

Figure 1.14 Scheme of supported liquid membrane with porous polymeric membrane.

the feed solution, as happens in extractive distillation using a third component. For this reason, the pervaporation process is considered as an alternative for the separation of liquid mixtures that are difficult or impossible to separate by conventional distillation.

1.4.6   *Vapour Permeation (VP)*

Vapour permeation is a membrane operation driven by a chemical potential difference across the membrane in which the feed stream, the retentate stream and the permeate stream are in the vapour phase. As in pervaporation, the permeate partial pressure is maintained by use of an inert sweep gas, vacuum or hygroscopic salt solutions.

1.4.7   *Liquid Membranes (LM)*

Liquid membranes (LMs) are traditionally divided into two categories. Those which contain only liquid phases are best represented by the double-emulsion liquid membrane (ELM) already patented two decades ago (Li, 1968). They are extensively described in the literature (e.g. Ho and Sirkar, 1992). In the water-in-oil-in-water (w/o/w) arrangement the feed usually forms the bulk aqueous phase, the extractant is the oil phase and the stripping solution is the inner aqueous phase (Kinugasa et al., 1990).

Liquid membranes of the second category involve polymeric membranes in addition to the liquid phases. The best-known example is the supported liquid membrane (SLM), in which a porous polymeric membrane, impregnated with the extractant, is situated between the feed and the stripping solution (Figure 1.14). Mass transfer models for this process are available in the literature (Chan and Lee, 1987; Chaudhuri and Pyle, 1992).

1.4.8 *Pertraction*

Pertraction or perstraction is an operation in which a multicomponent liquid feed flows on one side of a nonporous membrane. Different species in the feed establish different sorption equilibriums at the feed–membrane interface. Each dissolved liquid species then diffuses through the membrane material if there is a concentration gradient in the membrane. The dissolved molecules desorb from the membrane at the other side of the membrane into a different flowing liquid phase variously identified as the sweep liquid, purge liquid, extracting liquid or stripping liquid. This liquid provides the concentration gradient for permeation. The interface of this liquid and the membrane is the second interface at which solute species undergo partitioning again. The sweep liquid may or may not extract from the penetrating species. Thus, the process can combine selective permeation and extraction. When the phases are liquid, the process is called membrane-based solvent extraction. If we broaden the above definition by including gaseous or biphasic (gas + liquid) feed streams, with a partial pressure and/or temperature differences as the driving force, we can accommodate several operations based on dense or porous membrane contactors, such as membrane gas absorption, or membrane-based vapour recovery heat exchangers. In the latter a humid gas flow, the feed, is circulated on the face of a water-permeable membrane. A colder water flow is circulated on the other face of the membrane, and causes the condensation of water vapour on the feed side. The condensed water permeates through the membrane and is gathered in the cooling stream. The driving force of the separation is the temperature difference between the faces of the membrane. Net mass and heat flow from the feed to the permeate occur.

Pertraction can be realized through systems in which the extractant forms a thin liquid phase situated or flowing between two polymeric membranes rather than being immobilized on one. This mode of operation is also referred to as membrane-based solvent extraction (MBSX) (Wang and Bunge, 1990a, 1990b; Ho and Sirkar, 1992) and hybrid liquid membranes (HLMs) (Basu and Sirkar, 1991). Although various therms are utilized, membrane systems where the role of the membrane is just to optimize mass transfer between different phases, can be in general classified as membrane contactors.

1.4.9 *Membrane Distillation (MD)*

The process of membrane distillation makes use of hydrophobic porous membrane materials for separation. The material in this case is not wetted by the liquid feed and thus liquid penetration and transport across the membrane is prevented, provided that the feed side pressure does not exceed the minimum entry pressure for the pore size distribution of the particular material. Separation is by virtue of vaporization at the mouths of the pores and vapour transport through the pore network of the membrane. The membrane exerts little influence on the separation (fractionation) of the liquid, as the vapour–liquid equilibrium is not disturbed.

Direct-contact membrane distillation is a temperature-driven membrane operation in which the liquid on both sides of the membrane is in direct contact with the membrane and in which the liquid at the downstream side is used as a condensing medium.

Gas-gap membrane distillation is a system in which the downstream permeate side is condensed against a cool surface and where the condensed liquid at the downstream side is not necessarily in contact with the membrane; condensation takes place inside the module. The vapour pressure difference resulting from the temperature difference causes vapour molecules to be transported through the pores of the membrane (comparable to the evaporation, transport and condensation processes in distillation). The materials are typically polypropylene, PTFE and PVDF with submicrometre pore size, which have penetration pressures of several bars. These membranes are applied in other processes, such as microfiltration and as supports for liquid membranes.

1.4.10 Gas Separation

Gas separation is a pressure-driven membrane operation in which gas mixtures are separated by homogeneous, dense membranes or porous membranes. Homogeneous, dense membranes separate owing to differences of solubility and diffusivity. Porous membranes separate owing to differences in Knudsen diffusion and surface flow. Gas separation membranes are characterized by their permeability and selectivity. Asymmetric and composite membranes have been realized, as plane sheets and assembled in spiral wound modules or as hollow fibres.

Separation of air components, natural gas dehumidification, separation and recovery of CO_2 from biogas and of H_2 from industrial gases are examples of already existing large-scale industrial applications (Roman, 1996; Puri, 1996).

1.4.11 Electrodialysis (ED)

Electrodialysis is a molecular weight separation process in which electrically charged membranes and an electric potential difference are used to separate ionic species from an aqueous solution and other uncharged components. A typical electrodialysis arrangement consists of a series of anion and cation exchange membranes arranged in an alternating pattern between an anode and a cathode to form individual cells. A cell consists of a volume with two adjacent membranes. If an ionic solution such as an aqueous salt solution is pumped through these cells and an electric potential difference is established between the anode and the cathode, the positively charged cations migrate toward the cathode and the negatively charged anions toward the anode. The cations pass easily through the negatively charged membranes, but are retained by the positively charged membranes. The opposite applies for the negatively charged anions. The driving force for the ion transport in the electrodialysis process is the applied electrical potential between the anode and the cathode. The overall result is an increase in the ion concentration in alternate components, while the other components simultaneously become depleted.

Ion exchange membranes are ion exchange resins in film form: the cation exchange membranes contain negatively charged groups fixed to the polymer matrix, and the anion exchange membrane contains positively charged groups fixed to the polymer matrix. The selectivity of ion exchange membranes results from the exclusion of co-ions from the membrane phase. This type of selectivity is called Donnan exclusion. The membrane selectivity depends on the concentration of the fixed ions, the valency of the co-ions, the valency of the counterions, the concentration of the electrolyte solution and the affinity of the exchanger with respect to the counterions.

The most desirable properties of ion exchange membranes are high permeselectivity, low electrical resistance, good mechanical and form stability, and high chemical stability.

The matrix of ion exchange membranes consists of hydrophobic polymers such as polystyrene, polyethylene or polysulfone.

Since in electrodialysis the entire transfer of electrical charges is due to the transport of ions, the mass flux is directly proportional to the electric current, which is given by

$$i = F \, Z \, r_n \, J_n$$

where i is the current density, F the Faraday constant, Z the electrochemical valence and J the ion flux. The subscript n refers to the individual ion species.

The basic properties of the most common membrane operations are summarized in Table 1.6.

1.5 Mass Transport through Membranes

In membrane processes the transport properties control the permeability and the selectivity of the membranes. These processes being irreversible, they can be described using the methods of irreversible thermodynamics. A component i is moved through the membrane by its direct force and may also move due to coupled flow induced by the movements of other species:

$$J_i = L_{ii}x_i + \sum_{j \neq i} L_{ij}x \qquad (1.1)$$

↑	↑	↑
Flux of i	Flux due to direct force on i	Flux due to forces acting on other species

where x represents the driving force and coefficients L_{ij} are said to be *phenomenological coefficients or conductances*. The coefficients L_{ij} can be overall (membrane system) values or intrinsic (membrane only) values depending on the location of the driving force. The most general form of driving force is the chemical potential gradient that can be developed by specific components of driving force, such as ΔC, ΔP, and so on.

For homogeneous (dense) membranes, the general flux equation (which also includes electric potential gradient), known as the extended Nerst–Planck equation, is

$$J_i = D_i\left(\Delta C_i + C_i\Delta \ln \gamma_i + C_i\tilde{V}\frac{\Delta P}{\tilde{R}T} + Z_iC_iF_k\frac{\Delta \Psi}{\tilde{R}T} \right) \qquad (1.2)$$

↑	↑	↑	↑
Flux of i	Activity (concentration driving force)	Pressure driving force	Electric driving force

where the dimension of J_i is $[ML^{-2}T^{-1}]$ and

D_i = diffusivity of i $[L^2T^{-1}]$
C_i = concentration of i $[ML^{-3}]$
γ_i = activity coefficient of i
V_i = molar volume of species i $[L^3M^{-1}]$
P = pressure $[ML^{-1}T^{-2}]$
R = universal gas constant (8.31 J K^{-1} mol^{-1})
T = temperature $[T]$
Z_i = valency of species i
F_k = Faraday constant (9.65 × 10^4 C mol^{-1})
Ψ = electric potential $[L^2$ kJ C^{-1} T$]$

From (1.2) it is possible to develop simpler equations to describe the flux of solute or solvent through homogeneous membranes.

In membrane processes like reverse osmosis and ultrafiltration the driving force is the difference of hydrostatic pressure applied at the two surfaces of the membrane. To have a net flux of mass through the membrane, this difference has to be higher than the difference of osmotic pressure due to the difference of concentration between the two faces of the membrane.

For the membrane used in reverse osmosis, generally dense, the transport mechanism is given by the solution–diffusion model. Each component of the pressurized solution dissolves in the membrane until equilibrium conditions are achieved and diffuses through the membrane under a pressure and concentration gradient. The chemical composition and the molecular conformation of the membrane therefore have a considerable influence on the permeation and rejection properties.

The flux of a component i can be expressed as follows:

$$J_i = \frac{\partial_i C_i}{RT} \operatorname{grad} \mu i = \frac{\partial_i C_i}{RT} \left(\frac{\partial \mu i}{\partial C_i} \operatorname{grad} C_i + V_i \operatorname{grad} P_i \right) \qquad (1.3)$$

where

J_i = flux of i [ML^{-2}T^{-1}]

D_i = diffusivity of i [L^{2}T^{-1}]

C_i = concentration of i [ML^{-3}]

μ_i = chemical potential of i [kJ M^{-1}]

v_i = molar partial volume of i [L^{3}M^{-1}]

P_i = pressure of i [ML^{-1}T^{-2}]

R = universal gas constant (8.31 J K^{-1} mol^{-1})

T = temperature [T]

The membranes used in ultrafiltration processes reject only the chemical species of high molecular mass. These membranes are microporous and their pores have a size equal to or less than the molecules that are to be rejected. The rejection of solutes can also be influenced by the repulsive forces that act on them (Van der Waal's forces for the uncharged species and electrostatic repulsion for electrolytes and membranes with fixed charges).

The transport of species in pores is both diffusive and viscous.

The rejection and the velocity of ultrafiltration can be controlled by the formation, on the pressurized face of the membrane, of very thin layers of high molecular mass species.

The flow through these membranes is laminar and the permeability they show for water is higher than that of the membranes used for reverse osmosis. If the pores of the membrane are treated as cylinders with constant diameter and perpendicular to the membrane surface, the flux of a Newtonian fluid through the membrane is given by Hagen-Poiseuille's law:

$$J = \frac{\varepsilon r^2 \Delta P}{8 \mu \tau \, \Delta x} \qquad (1.4)$$

where

J = flow rate through the membrane [LT^{-1}]

r = mean pore radius [L]

ΔP = effective transmembrane pressure [ML^{-1}T^{-2}]

μ = viscosity of the fluid permeating the membrane $[ML^{-1}T^{-1}]$

Δx = length of the channel [L]

ε = surface porosity of the membrane

In (1.4) $\Delta P = \Delta P_t - \Delta \pi$, where $\Delta P_t = P_f - P_p$ (ΔP_t is the hydrostatic transmembrane pressure; P_f and P_p are the feed and permeate pressures, respectively); $\Delta \pi = \pi_f - \pi_p$, the difference in the osmotic pressure between feed (π_f) and permeate (π_p). An equation used to describe the permeate flux through the membrane is the Merten's equation:

$$J = K(\Delta P_t - \Delta \pi) \quad \text{where } K \text{ is the permeability coefficient as}$$
$$\text{defined in the equation 1.4} \left(K = \frac{\varepsilon r^2}{8\mu\tau} \right) \tag{1.5}$$

In most UF processes, the osmotic pressure of the retained solute is negligible and $\Delta P = \Delta P_t$ is normally used in (1.5).

1.5.1 *The Concentration Polarization Phenomenon*

For good efficiency of a membrane separation process it is necessary to control not only the transport phenomena in the membrane but also the transport of the different species in the solution, particularly at the interface. Concentration polarization, for example, produces lower selectivities and permeabilities than the theoretical values and can induce the effects of poisoning. This phenomenon is inevitable in membrane processes because the selective transport through the membrane causes accumulation or depletion in the fluid adjacent to the membrane. It consists of the building-up of concentration profiles of the retained solutes (molecules or colloids) in the bulk of the solutions and possible creation of particle layers along the membrane wall. The increased solute concentration on the membrane surface gives a very high osmotic pressure, with a consequent decrease in the driving force ($\Delta P_t - \Delta \pi$) and, thus, in the flux through the membrane. The convective permeation flux, which brings the 'material' to the surface of the membrane, is balanced by the back-diffusive transport of particles towards the solution bulk. This competition between convective and back-diffusive transports lead to a steady state of filtration. An appropriate change in the operating parameters, such as transmembrane pressure, axial flow rate and temperature, can reduce the concentration polarization phenomenon. The permeation flux, in fact, can be influenced by the operating conditions, as discussed briefly in the following.

- *Feed concentration*: The permeation flux decreases as the feed concentration increases owing to the earlier building-up of the polarization layer and to the viscosity increase of the solution.

- *Temperature*: Increasing the temperature, there is an increase in the permeation flux. This is due to the effect of temperature on fluid properties such as diffusivity and viscosity. Since diffusivity increases with temperature, mass transfer coefficient values will increase. Moreover, the viscosity of fluid decreases with temperature, so the permeate flow across the membrane will be easier.

- *Flow rate*: Higher fluid flow rates give higher mass transfer coefficients and, consequently, higher permeation flux. Moreover, agitation and mixing of the fluid reduce the thickness of the polarization layer owing to the removal of part of the accumulated solute.

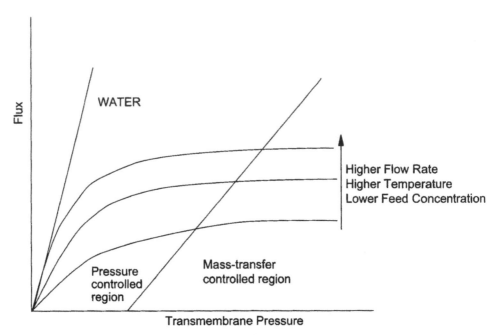

Figure 1.15   Typical behaviour of permeation flux as a function of transmembrane pressure.

- *Transmembrane pressure*: Typical behaviour of the permeation flux as a function of the transmembrane pressure is shown in Figure 1.15. At low pressures the curve is linear and the flux is directly proportional to the applied pressure. Increasing the transmembrane pressure, there is an increase of the solute concentration on the membrane surface and the permeation flux increases very slowly. At higher pressures, the flux becomes independent of pressure owing to consolidation of the polarized layer (mass transfer controlled region). Figure 1.16 summarizes the approaches that can be used to reduce the concentration polarization effects (Cheryan, 1986).

The main problem this phenomenon presents consists in the determination of the solute concentration at the membrane surface, on which the osmotic pressure, and then the permeation flux, depend. In the following, some of the main theories developed so far to describe the concentration polarization phenomenon for pressure-driven membrane operations are reported.

Film-theory   In the case of turbulent flow, the analysis of the system is based on the film theory, in which all the mass transport resistance is located in a film of liquid near the membrane and there is a uniform concentration in the bulk of fluid. Figure 1.17 is one illustration of this case.

If the mass transfer parallel to the membrane is neglected, the corresponding mass balance at steady state and abscissa x, assuming no solute passage, can be written as

$$J(x)\frac{\partial c}{\partial y} + D\frac{\partial^2 c}{\partial y^2} = 0 \tag{1.6}$$

where

$J(x)$ = solvent flux $[ML^{-2}T^{-1}]$
D = diffusivity $[L^2T^{-1}]$
C = solute concentration $[ML^{-3}]$

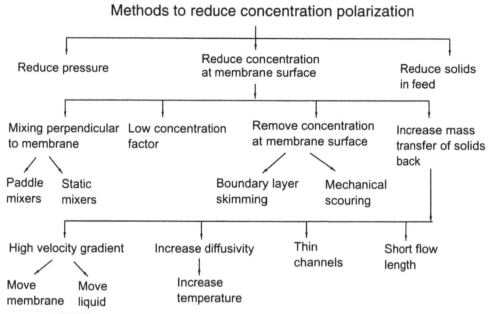

Methods to reduce concentration polarization

Figure 1.16   Methods for reducing concentration polarization effects.

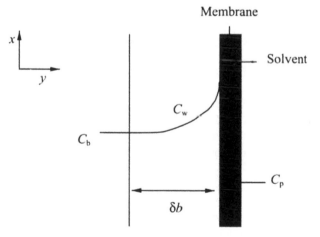

Figure 1.17   Schematic of concentration profiles according to the film theory.

Integration of (1.6) at constant x between $y = 0$ (membrane surface) where $C = C_w$ and $y = \delta b$ where $C = C_b$, gives

$$C_w/C_b = \exp[J(D/\delta b)] \tag{1.7}$$

| Polarization | Solvent | Solute mass transfer |
| modulus | flux | coefficient = K_s |

The left-hand side term represents the polarization modulus. K_s is the transport co-efficient of the particles towards the membrane wall; its value depends on physicochemical properties of the solution (viscosity, diffusion coefficient, etc.) and on the fluid dynamics of the system. Analogies with heat transport phenomena enable us to link K_s *empirically* with the different fluid dynamical parameters of the system.

A correction of the mass transfer coefficient for use in the film model, for both laminar and turbulent flow, is the Stewart correction θ_k (Bird et al., 1960):

$$\theta_k = \frac{K_s^*}{K_s} = \frac{V/K_s}{1 - \exp(-V/K_s)} \tag{1.8}$$

where K_s^* is the coefficient corrected for the effect of flux.

If convective transport is enhanced, C_w increases up to a maximum constant value C_g (gel concentration). At this point precipitation of the solute on the membrane surface or formation of a gel layer can be observed. In this case, the expression for the permeation flux is

$$J = K_s \ln(C_g/C_b) \tag{1.9}$$

An increase in the transmembrane pressure gives a thicker solute layer and, after an initial rise, the flux will drop back to the previous value. C_g and C_b are normally fixed by physicochemical properties of the feed; thus, the flux can rise only by increasing the K_s value. This can be achieved by enhancing the rate of removal of solute from the membrane wall.

The flux is calculated once the experimental determination of C_g and the empirical evaluation of the mass transfer coefficient have been made.

The film model provides good qualitative but unreliable quantitative descriptions of concentration polarization in membrane processes. It is impossible, in fact, to know the permeation flux value if there is no gel layer formation. Moreover, this approach is based only on a mass balance without taking into account the effects of the forces involved such as the applied transmembrane pressure. The basic equation used in this model has thus to be coupled to an equation in which the permeation flux is expressed as a function of the transmembrane pressure.

Song and Elimelech (1995) developed a novel theory in which equation (1.6) is coupled with a force balance. In this model C represents the excess concentration of particles in the solution. The authors defined the term M_p to describe the total number of excess particles accumulated per unit of membrane surface:

$$M_p = \int_0^\infty C \, dy \tag{1.10}$$

This expression is then coupled with the equation (1.6) and the flux conservation relationship to correlate the solute concentration and the permeation flux.

A relation between $V(x)$ and the operating parameters is obtained by a force balance in which the pressure drop across the layer is due to the drag force exerted by the retained particles on the permeate flow:

$$\Delta P_p = \int_0^{\delta b} FC \, dy \tag{1.11}$$

where

ΔP_p = pressure drop across the concentration polarization layer

F = drag force

The drag force is based on the Stokes–Einstein law modified by the Happel cell model (Einstein, 1956; Happel and Brenner, 1965):

$$F = \frac{KT}{D} A_S V(x) \qquad (1.12)$$

where

K = Boltzmann constant, 1.3803×10^{-23} J K^{-1}

T = temperature, K

D = diffusivity

A_S = correcting function introduced in Stoke's law:

$$A_S = \frac{1 + \frac{2}{3}\theta^5}{1 - \frac{2}{3}\theta + \frac{3}{2}\theta^5 - \theta^6}$$

$$\theta = (1 - \varepsilon)^{1/3}$$

To solve the equations involved in this theory, several hypotheses have to be made, such as complete particle retention ($C_p = 0$); Newtonian behaviour of the solvent flow; isometric, spherical and rigid particles. These assumptions allow the mathematical development of the theory. The calculation parameters and the resulting final equations are shown below.

Calculation parameters

$$N_F = \frac{4\pi a_p^3}{3kt} \Delta P_p$$

$$A_S(\theta^*) = \frac{N_F}{\theta_W^3}$$

$$\beta = \frac{kT(bA_S(\theta^*))}{D^2\gamma}$$

Final equations

$$V(x) = \frac{\Delta P}{(\lambda^3 + 6\beta \Delta P^2 x)^{1/3}} \left[\left(\sqrt{\frac{1^3}{\lambda^3 + 6\beta \Delta P^2 x}} + 4 + 2 \right)^{1/3} - \left(\sqrt{\frac{\lambda^3}{\lambda^3 + 6\beta \Delta P^2 x}} + 4 - 2 \right)^{1/3} \right]$$

$$V = \frac{\Delta P}{L\beta V(L)^2} \left(1 - \frac{\lambda}{\Delta P} V(L) \right)$$

$$C = \frac{\Delta P}{A_S(\theta^*)kT} \left[1 - V(x)\frac{\lambda}{\Delta P} \right] \exp\left(-\frac{V(x)y}{D} \right)$$

where

a_p = particle radius

k = Boltzmann constant, 1.3803×10^{-23} J K^{-1}

θ^* = value of θ between zero and θ_W

θ_W = value of θ at the membrane surface

C_b = bulk particle number concentration

D = particle diffusion coefficient

γ = fluid shear rate

ΔP = effective pressure drop (applied hydraulic pressure minus bulk osmotic pressure)

λ = membrane resistance

x = axial (longitudinal) coordinate

$V(x)$ = local permeate velocity (flux)

V = average permeate velocity (flux)

L = thickness of membrane

The main differences between the novel theory and the film theory are the following. In the film theory, the mass balance contains a transport coefficient that can be evaluated by means of empirical relations or experimental measurements; the basic equation has to be completed with semi-empirical equations to link the transport phenomena expressed in the mass balance to the driving forces, such as transmembrane pressure. In the novel theory, the mass balance equation is coupled with a second equation and it is possible to calculate permeate fluxes and concentration distribution and, thus, to describe the whole phenomenon of concentration polarization. However, the result is that it is more complex to use than the other formulation and has a more limited range of application, owing to the assumptions introduced.

Laminar boundary layer analysis   To obtain a more rigorous description of concentration and velocity profiles for conditions applying in membrane systems, it is necessary to apply the laminar boundary theory.

In this case, the mass transport resistance is not just located in a thin film of liquid near the membrane, but it involves all the fluid phase. This means that it is necessary first to determine the velocity profiles and then to solve the mass transport problem. The following equations are involved in the system:

Fluid continuity
$$\frac{\partial u}{\partial x} + \frac{\partial v}{\partial y} = 0$$

Solute continuity
$$\frac{\partial uc}{\partial x} + \frac{\partial vc}{\partial y} - D\frac{\partial^2 c}{\partial y^2} = 0$$

Momentum

(x direction)
$$u\frac{\partial u}{\partial x} + v\frac{\partial x}{\partial y} + \frac{1}{\rho}\frac{\partial P}{\partial x} - \nu\left(\frac{\partial^2 u}{\partial x^2} + \frac{\partial^2 u}{\partial y^2}\right) = 0$$

(y direction)
$$u\frac{\partial u}{\partial x} + v\frac{\partial v}{\partial y} + \frac{1}{\rho}\frac{\partial P}{\partial y} - \nu\left(\frac{\partial^2 v}{\partial x^2} + \frac{\partial^2 v}{\partial y^2}\right) = 0$$

where

u = velocity in axial (x) direction [LT^{-1}]

v = velocity normal to membrane (y direction) [LT^{-1}]

ν = kinematic viscosity [ML^{-1}T^{-1}]

P = pressure [ML^{-1}T^{-2}]

D = diffusivity [L^{2}T^{-1}]

ρ = density [ML^{-3}]

The above set of equations has been used in numerous cases. Solution involves further simplification and careful choice of boundary conditions. This method is more rigorous but often difficult to apply and it is limited to laminar flow.

The interaction of chemical species and their deposition on or within the membrane surface might be at the origin of irreversible transmembrane flux decay and selectivity changes. The phenomenon generally known as fouling can be induced by the increase of interfacial concentrations due to concentration polarization.

The overall increase of the mass transport resistance includes:

- adsorption of solutes to the membrane both within the pores and at the surface;
- partial and complete blocking of pores;
- deposition and adhesion of suspended material onto the membrane surface;
- consolidation of the concentration polarization layer, by the chemical interaction or physical compaction.

This leads to:

- reduction of the membrane surface;
- increase of the mass transport resistance.

The overall resistance is composed of a membrane resistance, which is assumed to remain unchanged, and a cake resistance.

Particles might enter the pores and either be deposited or be adsorbed, reducing pore volume. The irregularity of pore passages causes the particle to become tightly fixed blinding the pore. In this case, membrane resistance increases as a consequence of pore size reduction or of reduction of the membrane surface by complete or partial pore blocking. (Solid particles may also bridge a pore by together obstructing the entrance but not completely blocking it.)

Accounting for mechanisms of foulant removal from the membrane surface, constant-pressure dead-end filtration equations (Hermia, 1982) have been modified to describe crossflow microfiltration (Field et al., 1995) as follows:

$$-\frac{dJ}{dt} = k(J - J_{\text{lim}})J^{2-n}$$

where

J_{lim} = limit value of the permeate flux

k, n depend on the fouling mechanism:

Complete pore blocking: $n = 2$

Partial pore blocking: $n = 1$

Cake filtration: $n = 0$

Internal pore blocking: $n = 1.5$

1.5.2 *Methods for Reducing Fouling*

Concentration polarization and fouling can be considered intrinsic phenomena of particular significance in pressure-driven membrane operations. The methods for reducing the decline in transmembrane flux and in some cases the changes in membrane selectivity due to them can differ depending on the specific characteristics of the system. Studies on their prevention, by experimentally identifying optimal operating conditions or by reducing the concentrations of the species that might create the fouling, are very important. High axial flow rate combined with low transmembrane pressure (TMP), operating at constant permeate flux (instead of constant TMP) and below the critical flux, optimal module configuration, and the presence of appropriate turbulence promoters, can contribute to the control of the phenomenon. The nature of the membrane material for avoiding adsorption of the solutes, the pore size and the pore morphology are other parameters to be optimized case by case. Hydrophilic polymers are less fouling in aqueous environments. The use of non-ionic surfactant adsorbed on the pressurized membrane surface might reduce fouling phenomena. Adsorption is also important in small units and with valuable bioactive species in that it affects their recovery. Fouling can be controlled by immobilizing bio-catalysts (proteases, pectinases, etc.) which hydrolyzes foulants (such as proteins, pectins, etc.) as they accumulate on the membrane surface.

Membrane cleaning can be performed using specific chemicals that do not destroy the membrane materials, for the appropriate times and temperature and in general with very low or zero transmembrane pressure applied. Acids (citric, phosphoric, etc.) or alkali (NaOH), detergents, enzymes, and complexing agents have been tested successfully. Mechanical cleaning by introducing soft sponge balls in the tubular membranes can also be utilized.

Microfiltration and some ultrafiltration membranes can also be cleaned by a hydraulic method called back-flushing, reversing the direction of the permeation across the membrane itself. This is achieved by releasing the feed pressure and applying pressure on the permeate side. A liquid (the permeate itself) or a gas is generally transported across the membranes from the permeate side to the feed side, contributing to the mechanical removal of the substances deposited. The operation can be repeated automatically with a determined frequency to maintain the transmembrane flux at optimal values. Other techniques such as pulsatile flow, vortex formation, and continuous or intermittent electrical pulses have been also suggested.

1.6   Industrial Applications of Membrane Operations

Membrane operations have already been applied successfully in a large variety of industrial and medical processes. Most of them are recognized as basic unit operations in chemical engineering. The design of integrated membrane systems is also becoming attractive where the molecular separations are combined with chemical conversions in membrane reactors and with other traditional separation units.

Loeb and Sourirajan, with their discovery in the early 1960s of how to increase significantly the permeability of CA polymeric membranes without significant changes in their selectivity, made real the possibility of using membranes in large-scale operations for desalting brackish water and sea water and for various other molecular separations in different industrial areas.

Millions of cubic metres of desalted water are produced daily in the world by reverse osmosis; hundreds of plants with individual capacities of the order of 10^5 m^3 per day are in operation.

In Table 1.8a and 1.8b is reported the growth in reverse osmosis desalination in the 10 years to 1992 and of water production worldwide. It is interesting to note the high production of reverse osmosis ultrapure water in Japan, considering the strategic role of the electronic industry in that country.

Billions of cubic metres of pure gases are also produced via selective permeation in polymeric membranes. More than 10 000 units have been already installed, for example, for production of pure nitrogen from air.

Table 1.8 (a) Percentage growth in use of reverse osmosis in water purification (1982–1992) and (b) Percentage uses of RO in water production (1992)

	Japan	World
(a)		
Brackish water	34	67
Sea water	3	18
River water	–	4
Pure water	–	6
Waste water	8	4
Tap water	44	–
Unknown	10	–
Others	1	1
(b)		
Industrial water	3	39
Potable water	2	36
Boiler feed water	12	10
Discharge	–	9
Military	–	2
Tour	–	2
Ultrapure water	79	–
Reuse	3	–
Others	1	2

Source: Water Reuse Promotion Center, Japan (1992).

Membrane systems are becoming competitive with most of the traditional cryogenic and PSA (pressure swing absorption) units.

Pressure-driven membrane operations (crossflow microfiltration, ultrafiltration, nanofiltration, reverse osmosis) have been utilized by process engineers for innovative production processes in various areas such as the dairy, textile, leather and agrofood industries to minimize environmental problems and energy consumption with a more rational utilization of raw materials and recovery and reuse of by-products.

The newest membrane operations such as pervaporation and membrane contactors are growing rapidly or are under test at semi-industrial level (e.g. for osmotic distillation and membrane distillation). Hundreds of thousands of square metres of UF and RO membranes have been installed in recent years in the dairy industry (see Figure 1.18) for the treatment of whey and in general for redesigning production lines.

A large number of studies have been carried out in the agrofood industry, where ultrafiltration and reverse osmosis units have been introduced in various processes. The food industry traditionally requires large amounts of thermal energy, particularly in the sterilization, concentration and drying of the products. Reverse osmosis is the membrane technology that is substituting concentration by evaporation in various industrial plants.

From energy consumption analysis, it appears that evaporation requires the largest amount of thermal energy compared to the other steps. As an example, the cycle for concentrating tomato juice from 5.5 ° Brix to 29 ° Brix can be modified by using reverse osmosis membranes (Drioli et al., 1988). The overall thermal energy is decreased owing to the possibility of heat recycling because the reverse osmosis unit is fed at much lower temperature. To overcome (i) the high osmotic pressure deriving from the high concentration level of

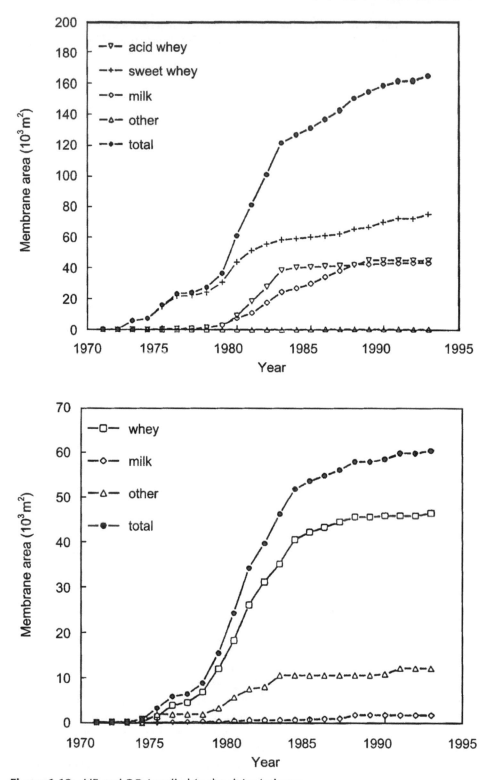

Figure 1.18 UF and RO installed in the dairy industry.

the treated juice, (ii) the fouling of the membranes, and (iii) the concentration polarization and the increased viscosity during the process, an integrated membrane process can be used. In principle it is possible to make a first gross separation between the pulp and the aqueous solution containing low molecular mass compounds such as mineral salts, vitamins, etc. This can be achieved by means of an appropriate choice of a microfiltration (MF) membrane unit positioned after the pulper and finisher. The serum is then concentrated in a typical reverse osmosis unit that operates under much better conditions than in the previous cycle. The two outcoming fluid streams are then mixed together to reconstitute the product.

A similar approach has already been applied at industrial level for the concentration of citrus fruit juices. A membrane process named Fresh Note™ (FMC Corp. and E.I. Du Pont) combines ultrafiltration, reverse osmosis, pasteurization and blending to produce a concentrate juice of very high quality. Successful processing of citrus requires the treatment of some compounds to ensure their stability, while other compounds have to be kept cold to preserve quality. The UF process allows separation of critical flavour compounds from the elements that must be heat pasteurized to ensure overall product stability. When fed with single-strength juice, the UF produces two streams. The clarified permeate serum, which represents about 95% of the original volume, is sterile and contains almost all of the delicate flavour compounds. The remaining 5% retentate pulp stream contains all of the suspended solids, bacteria, pectins, moulds and yeasts. This stream is quickly heated, then cooled to ensure final product stability. The serum is concentrated in the reverse osmosis step at about 62 ° Brix, operating with appropriate membranes at about 102 bar. Flavour molecules are retained in the concentrated stream. The blending process combines the pasteurized pulp stream with the concentrated serum to produce a final juice at 42–51 ° Brix.

The possibility of rationalizing the wine industry with the use of integrated membrane systems, particularly when large-scale production of high-quality wine is required, is also attractive (Drioli and Molinari, 1990). In Figure 1.19 a suggested cycle is presented. Grape musts can be stabilized and clarified immediately after the pressing by crossflow microfiltration or ultrafiltration, without changing the sugar content or the fraction of low molecular mass polyphenols and controlling the amount of high molecular mass polyphenols. The must obtained will be an ideal substrate for the action of selected yeasts. Reverse osmosis or membrane distillation can be used when necessary to concentrate the must to an optimum sugar content before fermentation.

The fermentation itself might take place in an enzyme membrane reactor, followed by ultrafiltration to separate and clarify the produced wine. Reverse osmosis or technical dialysis might also be introduced downstream to reduce the alcohol content of the wine produced if this is of interest.

Electrodialysis can be applied for the partial removal of ions when required. Crossflow microfiltration, ultrafiltration and reverse osmosis have recently been considered by the Office International de La Vigne et du Vin in the new recommendation on the technologies permitted in wine making; however, reverse osmosis was only considered at experimental level.

Dyeing of textiles demands large quantities of water, and also results in large amounts of wastewater streams from the different steps, with some of them containing high loads of chemical oxygen demand (COD) and dye. The main problem is represented by the residual dyes which, before discharging, require degradation. The latter is normally very low (owing to the dilution), forcing industry to install tertiary treatments of increasing capacity to reduce COD and colour to levels acceptable by law. An additional problem might be generated by salts of various nature dissolved in the effluent water. The total

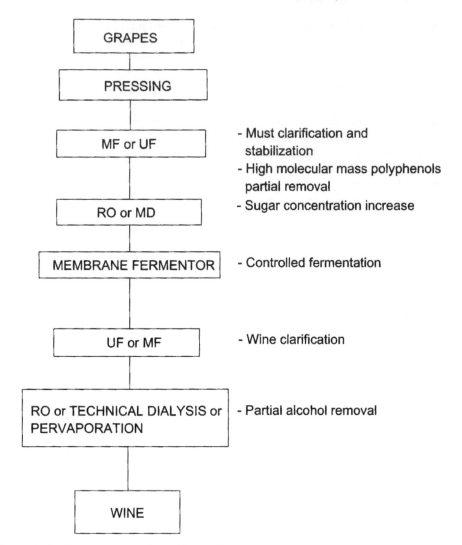

Figure 1.19 Processing of wine with membrane processes.

water consumption can vary from 50 to 200 litres of fresh water per kilogram of textile, depending on the type of equipment and whether it is a continuous or batchwise process. In addition, the COD amounts to 4000 mg l^{-1} and the dye up to 70 mg l^{-1}, depending on the type of effluent stream (Molinari et al., 1995).

Membrane processes can be used for the treatment of final wastewater and for recycling warm water, dyes and auxiliary chemicals within the cycle. In fact, membrane filtration minimizes the consumption of dyeing process chemicals (change in recipes), and saves energy through the recycling of effluent streams (usually hot at 90 °C) that are currently allowed to drain away, reducing energy consumed in heating up fresh water.

Through proper pretreatment, followed by nanofiltration or reverse osmosis, it has been possible to concentrate waste effluents from textile dye-houses by a factor of 5 to 9, and obtain pure water equivalent to 50–85% of the same stream; this pure water is recycled to the dyeing process. The concentrated solution is recycled to biological treatment or is post-treated according to established processes.

Table 1.9 Technical data for industrial units using composite membranes

Code	Location	Membrane area (m^2)	Solvent	Flow rate (kg h^{-1})	Concentration		Consumption	
					Inlet (%)	Outlet (%)	Steam (kg h^{-1})	Power (kW)
PVA1	Betheniville, France	2100	Ethanol	5000	96	99.8	560	200
				3000	96	99.95	500	
PVA2	Provins, France	480	Ethanol	1195	85.7	99.8	195	85
				1500	93.9	99.8	110	
				840	85.7	99.95	145	
				970	93.9	99.95	83	
PVA3	ICI, Australia	210	Isopropanol	500	96	99.9	30	−10
PVA4	BASF, USA	60	Solvenon	75	55	98	204	5
PVA5	Nattermann, Germany	28	Ethanol	2600	96	96.4	70	7
PVA6	ALKO, Finland	12	Ethanol	19	94.5	99.8	–	11
PVA7	Nattermann, Germany	80	Ethyl acetate	250	96.5	99.8	25	5

The separation and recovery of organic solvents and dehumidification of gas streams is also growing rapidly at industrial level (Baker, 1985). In 1989 the first vapour recovery unit (VRU), based on membrane technology, was commissioned for off-gas treatment in a gasoline tank farm. At present 20 membrane VRUs are in operation or in construction. The capacity of these units ranges from 100 to 2000 m^3 h^{-1}. These are single membrane stages or hybrid systems of a membrane stage combined with a post-treatment facility, for example a catalytic incinerator, gas engine or pressure swing adsorption. These plants are equipped with GKSS flat sheet membranes and the GKSS 'GS-module'. The GS-module is a modified plate-and-frame configuration. A standard module is 500 mm long and has a diameter of 320 mm. The installed membrane area per module varies from 8 to 10 m^2.

Volatile organic solutes may be efficiently retained by reverse osmosis membranes; however, non-ionized molecules are not effectively rejected, but the elimination of volatile solutes from dilute aqueous solutions may be possible by pervaporation.

A possible application of the removal of organic solutes could be the treatment of industrial and municipal water supplies contaminated with carcinogenic halogenated organic compounds. Such a process would also be attractive for the extraction of organics (ethanol, acetone, etc.) produced by fermentation. The separation of aqueous organic mixtures by pervaporation membranes, particularly to remove traces of hydrocarbons, has also been studied. High selectivity was generally observed for aromatic and chlorinated hydrocarbons.

Industrial companies such as Deutsche Carbone (GFT), BP International (Kalsep), Lurgi, Sempas, MTR, Separex, Mitsui, Daicell, and Tokuyama Soda, are taking part in industrial applications of the pervaporation process. The membranes used in the equipment are generally in composite form, with the selective layer made of modified poly(vinyl alcohol). The technical data for some units are summarized in Table 1.9 (Zhang and Drioli, 1995).

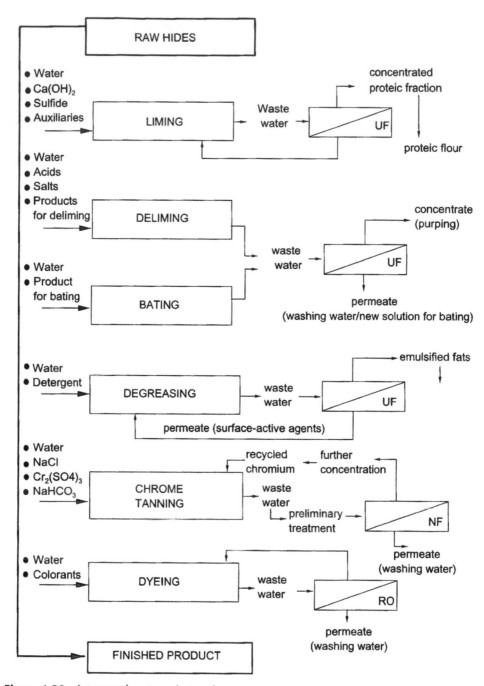

Figure 1.20 Integrated systems in tanning processes.

The tanning process, traditionally characterized by a low technological content, can be completely redesigned as a biotechnological process by introducing various enzymatic steps and molecular separations. Research programs are in progress today, particularly in Italy, where more than two thousand tanneries are in operation. The tentative innovative process is represented in Figure 1.20 and compared with a traditional one. Studies for the

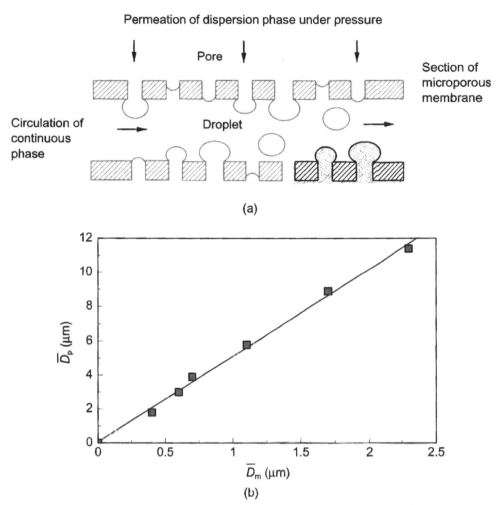

Figure 1.21 (a) Production of microemulsions by permeation through microporous membranes. (b) Relationship between droplet size (D_p) and membrane pore diameter (D_m) (Furuya et al., 1996).

development of membrane contactors and pertraction processes are also contributing to new industrial applications. The possibility of producing monodispersed microemulsions by permeation in hydrophobic or hydrophilic microporous membranes has been studied (Furuya et al., 1996) and realized industrially (Figure 1.21a). An interesting relationship between droplet size and membrane pore diameter has been observed (Figure 1.21b).

Membrane contactors can be utilized for developing phase transfer catalysis. A phase transfer catalyst is a substance that increases the rate of reaction between substrates present in separate phases. The use of these catalysts is growing owing to their ability to promote chemical transformations that take place in polar organic solvents, reducing the necessity for mutual solvents. The possibility of using membrane-supported phase transfer catalysis in which a membrane separates the organic and aqueous phases while promoting reaction between them is particularly attractive (Taverner and Clark, 1997). At its best the membrane reactor can realize an ideal surface for the reaction at the phase boundary.

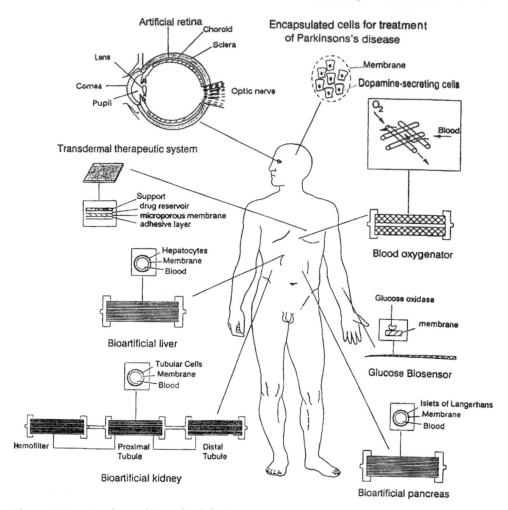

Figure 1.22 Membrane biomedical devices.

Membrane biomedical devices already in use are summarized in Figure 1.22. The integrated membrane systems for industrial production already developed in industrial cycles might be applied also in the design of integrated artificial organs.

The instances described in these pages are only examples of the potential membrane operation systems. Practically every industrial separation problem and potentially any important conversion process might be innovated by introducing membrane technology.

1.7 References

BAKER, R.W., 1985, Process for recovery of organic vapors from air, U.S. Patent 4,533,983, Nov. 19.

BASILE, A., CRISCUOLI, A., SANTELLA, F. and DRIOLI, E., 1996, Membrane reactor for water gas shift reaction, *Gas Sep. Purif.*, **10**(4), 243–254.

BASU, R. and SIRKAR, K.K., 1991, Hollow fibre contained liquid membrane separation of citric acid, *AIChE J.*, **37**, 383–393.

BIRD, R.B., STEWART, W.E. and LIGHTFOOT, E.N., 1960, *Transport Phenomena*, Wiley, New York.

BOOM, S.P., PÜNT, I.G.M., ZWIJNENBERG, H., DE BOER, R., BARGEMAN, D., SMOLDERS, C.A. and STRATHMANN, H., 1998, Transport though zeolite filled polymeric membranes, *J. Membr. Sci.*, **138**, 237–258.

BROCK, T.D., 1983, *Membrane filtration*, Science Tech., Inc., Springer-Verlag, Berlin.

CADOTTE, J.E. and PETERSON, R.J., 1981, Thin film composite reverse osmosis membranes: origin, development and recent advances, in *Synthetic Membranes*, Vol. 1, *Desalination*, pp. 305–326. ACS Symp. Series no. 153.

CHAN, C.C. and LEE, C.J., 1987, A mass transfer model for the extraction of weak acids/bases in emulsion liquid-membrane systems, *Chem. Eng. Sci.*, **42**(1), 83–95.

CHAUDHURI, J.B. and PYLE, D.L., 1992, Emulsion liquid membrane extraction of organic acids. I. A theoretical model for lactic acid extraction with emulsion swelling, *Chem. Eng. Sci.*, **47**(1), 41–48.

CHERYAN, M., 1986, *Ultrafiltration Handbook*, Technomic Publishers, New York.

DRIOLI, E. and MOLINARI, R., 1990, Membrane processing of must, wines and alcoholic beverages, *Chimica Oggi*, 47–55.

DRIOLI, E., CALABRÒ, V., MOLINARI, R. and DE CINDIO, B., 1988, An exergetic analysis of tomato juice concentration by membrane processes, in S. BRUIN (ed.) *Preconcentration and Drying of Food Materials*, pp. 103–114, Elsevier, Amsterdam.

EINSTEIN, A., 1956, *Investigation on the Theory of the Brownian Movement*, R. Furth ed., Dover, New York.

FIELD, R.W., WU, D., HOWELL, J.A. and GUPTA, B.B., 1995, Critical flux concept for microfiltration fouling, *J. Membr. Sci.*, **100**, 259–272.

FRENANDER, U. and JÖNSSON, A.-S., 1996, Cell harvesting by crossflow microfiltration using a shear-enhanced module, *Biotechnol. Bioeng.*, **52**, 397–403.

FURUYA, A., ASANO, Y., KATOH, R., SATOYAMA, K. and TOMITA, M., 1996, Preparation of food emulsions using a membranes emulsification system, *Proc. ICOM '96*, Yokohama, p. S-14-2-1.

GEUS, E.R. and VAN BEKKUM, H., 1989, A ZSM-5 crystal membrane model, *Proc. 8th IZC*, Amsterdam, pp. 293–295.

GEUS, E.R., MULDER, A., VISHJAGER, D.J., SCHOONMAN, J. and VAN BEKKUM, H., 1991, *ICIM '91*, pp. 57 64.

HAPPEL, J. and BRENNER, H., 1965, *Low Reynolds Number Hydrodynamics*, Prentice-Hall, Englewood Cliffs, NJ, pp. 6–30.

HENIS, J.M.S. and TRIPODI, M.K., 1981, *J. Membr. Sci.*, **8**, 233.

HERMIA, J., 1982, *Trans. IChemE*, **60**, 183–187.

HO, W.S.W. and SIRKAR, K.K., 1992, *Membrane Handbook*, Van Nostrand Reinhold, New York.

JIA, M., CHEN, B., NOBLE, R.D. and FALCONER, J.L., 1994, Ceramic-zeolite composite membranes and their application for separation of vapor/gas mixtures, *J. Membr. Sci.*, **90**, 1.

KESTING, R.E., 1971, *Synthetic Membranes*, McGraw-Hill, New York.

KINUGASA, T., WATANABE, K. and TAKEUCHI, H., 1990, Stability of (w/o) emulsion drops and water permeation through its liquid membrane in (w/o)/w dispersion, *Proc. Int. Congr. Membranes and Membrane Processes*, Chicago, Vol. I, pp. 706–708.

KOOPS, G.H., 1995, *Nomenclature and Symbols in Membrane Science and Technology*, European Society of Membrane Science and Technology.

LI, N.N., 1968, Separation of hydrocarbons with liquid membranes, U.S. Patent 3,410,794.

LIU, Q., NOBLE, R.D., FALCONER, J.L. and FUNKE, H.H., 1996, Organic/water separation by pervaporation with a zeolite membrane, *J. Membr. Sci.*, **117**, 163–174.

LOEB, S., 1993, The Loeb–Sourirajan membrane: how it came about, *Desalination & Water Reuse*, **3**(2), 20–24.

LOEB, S. and SOURIRAJAN, 1962, *Adv. Chem. Ser.*, **38**, 117.

MOLINARI, R., GAGLIARDI, R. and DRIOLI, E., 1995, Methodology for estimating saving of primary energy with membrane operations in industrial processes, *Desalination*, **100**, 125–137.

PURI, P., 1996, Membranes for gas separation: current status, in *Proc. Seminar on the Ecological Applications of Innovative Membrane Technology in the Chemical Industry*, Cetraro, p. C17.

PUSCH, M. and WALCH, A., 1982, *Chem. Ind., Ed. Engl.* **21**, 660–685.

ROMAN, I.C., 1996, Membranes for N$_2$ generation: recent advances and impact on other gas separations, in *Proc. Seminar on the Ecological Applications of Innovative Membrane Technology in the Chemical Industry*, Cetraro, p. C16.

ROZELLE, L.T., KOPP, C.V., CADOTTE, J.E. and COBIAN, K.E., 1975, *Trans ASME, J. Engl. Ind.*, 220–223.

SCHNEIDER, K., 1981, *Kunstoffe*, **71**, S183–184.

SONG, L. and ELIMELECH, M., 1995, Theory of concentration polarization in crossflow filtration, *J. Chem. Soc. Faraday Trans.*, **91**(19), 3389–3398.

TAVERNER, S.J. and CLARK, J.H., 1997, Recent highlights in phase transfer catalysis, *Chem. Ind.*, 22–27.

VAN HENVEN, J.W. and BLOEBAUM, R.K., 1974, RO by dynamically formed cation exchange membranes, *Desalination*, **14**, 229.

WANG, C.C. and BUNGE, A.L., 1990a, Multisolute extraction of organic acids by emulsion liquid membranes. I. Batch experiments and models, *J. Membr. Sci.*, **53**(1–2), 71–103.

WANG, C.C. and BUNGE, A.L., 1990b, Multisolute extraction of organic acids by emulsion liquid membranes. II. Continuous flow experiments and models, *J. Membr. Sci.*, **53**(1–2), 105–126.

ZHANG, S.M. and DRIOLI, E., 1995, Pervaporation membranes, *Sep. Sci. Technol.*, **30**(1), 1–31.

For further information on the various topics presented in these pages the following (books) are suggested

DRIOLI, E. and NAKAGAKI, M. (eds), 1986, *Membranes and Membrane Processes* (Proceedings of Europe-Japan Congress on membranes and membrane processes, Stresa, Italy, June, 1984), Plenum Press, New York.

HO, W.S.W. and SIRKAR, K.K., 1992, *Membrane Handbook*, Van Nostrand Reinhold, New York.

HSIEH, H.P., 1996, *Inorganic Membranes for Separation and Reaction*, Elsevier Science, Amsterdam.

ZEMAN, L.J. and ZYDNEY, A.L., 1996, *Microfiltration and Ultrafiltration*, Marcel Dekker, New York.

2

Catalytic Membrane Reactors (CMRs)

2.1 Introduction

In contrast to inorganic catalysts, biocatalysts such as enzymes and cells feature selective and effective catalytic action that permits them to perform under mild conditions of pH, pressure and temperature reactions that would otherwise require drastic operating conditions. Since the advent of what has been called solid phase biochemistry, the advantages of immobilized biocatalytic preparations for homogeneous phase enzymatic/cellular reactions have been exploited to develop new and less expensive processes.

Synthetic membranes provide an ideal support for immobilization of the biocatalyst, owing to a large available surface per unit volume, and potential for the development of new immobilization procedures. Immobilized enzymes are retained at the reaction site, do not pollute the products and can be continuously reused. Substrate partition at the membrane/fluid interphase can be used to improve the selectivity of the catalytic reaction towards the desired products with minimal side reactions (Trevan, 1981).

Although, in a specific reaction, enzymes play a role similar to that of synthetic catalysts, leading to a substantial reduction of reaction activation energy, their catalytic action is extremely efficient and selective, well beyond the performance of synthetic catalysts. It is not surprising, therefore, that the prospect of their application has prompted an enormous interest both in academic research and in industry, leading to a series of related developments such as enhancement of the techniques for inducing microorganisms to produce selected enzymes; improved enzyme purification techniques; engineering of techniques to immobilize enzymes or whole cells on or in solid supports; development of techniques for enzyme usage in continuous flow reactors; and development of techniques for the solid phase synthesis of peptides.

Enzymes produced by microorganisms such as fungi or bacteria have been used for years in batch fermentation plants for the production of pharmaceuticals, beverages, foods, and so on. The availability of almost pure enzymes enables one to carry out specific reactions under mild conditions, to limit side product formation and to perform syntheses of chemically active compounds which would otherwise require extremely long reaction times with low yields. The use of enzymes as biocatalysts can therefore be of extreme interest for individual applications mainly for the advantages it implies in terms of energy consumption, safety, pollution prevention, and materials preservation.

The traditional use of enzymes in stirred batch and similar reactors is limited, however, for the following reasons: high enzyme purification costs; low productivity per reactor, per unit time; difficult and expensive recovery and reuse of enzymes or cellular microorganisms; product pollution; and difficulties in maintaining standard product quality (Michaels, 1968; Drioli and Catapano, 1984).

Most of the above problems could be overcome by using 'immobilized enzymes', in other words enzymes confined or compartmentalized in a well-defined region of space, which retain their catalytic properties and can be repeatedly and continuously used. Several advantages can derive from the use of immobilized enzymes or microbial cells, including enzyme reuse; product separation; enhanced enzyme stability; enhanced enzyme stereoselectivity; reduction of operating costs; opportunity to design processes in a more rational way; more compact plants; high productivity per unit time per item of equipment, with small amounts of side products.

A lot of research work has therefore been performed to optimize immobilization techniques and procedures in view of the development of enzyme reactor engineering (Mosbach, 1980; Atkinson and Mavituna, 1984; Telo et al., 1990; Dalvie and Baltus, 1992). Membrane science contributes to the development of immobilized enzyme reactors for industrial processes. This well-assessed technology in fact seems very promising for the design of compact and flexible apparatus and particularly useful when a separation step is to be coupled to the chemical reaction.

Insoluble carrier-fixed enzymes can be used continuously in stirred vessels or tubular reactors without catalyst loss, and can then be recovered by means of sedimentors, magnetic fields or similar apparatus. On the other hand, in nature a continuous uptake of substrate and release of product without loss of biocatalysts is achieved not by carrier fixation but by means of cellular membranes. Efficient immobilized enzyme reactor systems for technical applications can therefore be established replacing cellular membranes by synthetic membranes, and the activated transport through the cellular wall by a forced transport across the membrane (Wandrey, 1983). Traditional immobilization techniques are summarized in Figure 2.1.

Ultrafiltration membranes can be used to retain enzymes in the reaction vessel owing to size difference between the usual high molecular mass enzymes and the mostly low molecular mass substrates and products; homogeneous continuous catalysis can thus be achieved. Some membrane reactor configurations do not allow a homogeneous distribution of catalyst in the vessel, so that the reactor acts in a more heterogeneous way. Depending on enzyme kinetics, different reactor configurations exist that are able to optimize system yield; systems in which product concentrations are kept low and continuous product removal occurs are particularly interesting when dealing with product-inhibited enzymes (Leuchtenberger et al., 1984).

Catalytic systems in which membranes are used as separation media and/or immobilization support (and not as catalyst carriers) are called 'catalytic membrane reactors' (CMR). The most general configurations are shown in Figure 2.2. Enzymes are the most commonly used biocatalysts, and in this case the catalytic membrane reactor is termed an enzyme membrane reactor (EMR). The kinetics of immobilized enzymes are widely discussed in the literature (Bailey and Ollis, 1986; Greene, private communication, 1983) and are reported in Appendix 2.1.

As well as enzymes, microbial, plant and animal cells, catalytic antibodies (or abzymes) and catalytic RNA (or ribozymes) are also used as biocatalysts.

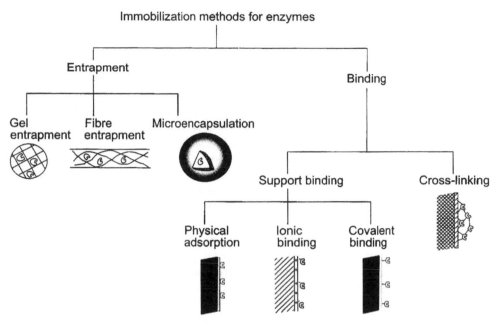

Figure 2.1 Classification of immobilization methods for enzymes (ɕ = enzyme).

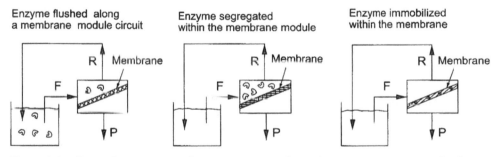

Figure 2.2 General enzyme membrane reactor configurations (ɕ = enzyme; F = feed; R = retentate; P = permeate).

Concentration polarization phenomena severely affect the performance of reactors when the membrane is used only as a separation medium, so that it is necessary to control the polarization layer on the membrane pressurized side by means of reactor fluid dynamics or appropriate design. Fluid dynamic conditions in some of these reactors make them especially suitable for enzymatic systems for which a homogeneous catalyst distribution is particularly important, such as cofactor-requiring mono- and multienzyme systems.

Concentration polarization phenomena, which are the main drawbacks of the aforementioned enzyme membrane reactors, can nevertheless be used to form, either in dynamic or in static conditions, a gel layer of enzyme proteins on a membrane. It is even possible to establish more than one enzyme layer and no coupling agent is needed to carry out the immobilization. Owing to high protein concentration on the membrane surface, enzyme stability can be improved over that in systems using enzymes homogeneously distributed in the reacting solution. Reduced catalytic efficiency due to mass transport limitations and the possibility of preferential pathways in the enzyme gel layer can be typical system disadvantages (Drioli and Scardi, 1976).

Asymmetric hollow-fibre membranes can also be used as selective supports for enzymes. A biocatalyst suspension can in fact be forced through the unskinned surface of asymmetric membranes so that biocatalysts, either enzymes or whole cells, although still suspended, are effectively immobilized within the macroporous spongy part of the membranes. The enzymatic activity can thus be spread over a large surface, although substrates and products can only diffuse to and from the biocatalyst.

In certain applications, the size of the biocatalyst is not suitable for entrapment into the supporting sponge layer of asymmetric membranes. In these cases, the membrane acts mainly as a selective barrier; it defines a reactive zone in the vessel, usually the shell, accessible to substrates and products mainly by diffusive mass transfer, preventing the catalyst from pollution or inhibition that might be caused by other species in solution.

Enzymes can still be absorbed within symmetric macroporous membranes in order to establish high catalyst concentrations, cross-linked to prevent elution, or simply covalently or ionically bound to either symmetric or asymmetric membranes. In spite of short residence times, high conversions can be achieved in most kinds of enzyme membrane reactors (Howell et al., 1978; Horvath et al., 1973; Adu-Amankwa and Constantinides, 1984). Finally, viable cells are grown and used in membrane fermentors for a wide range of applications (Porter and Michaels, 1970; Mehaia and Cheryan, 1984; Shimizu et al., 1993).

Important parameters that must be considered in immobilized enzyme reactors are enzyme activity; half-life time and activity decay profile; optimal substrate concentration; optimal residence time; pH and temperature; by-product formation; inhibitors; pressure drop; mode of flow; and particle size, shape and distribution.

The different main configurations of catalytic membrane reactors will be discussed in the following pages. Systems in which the biocatalysts are bound or entrapped in the membranes (forming catalytic membranes) or confined in a well-defined region of the reaction vessels, and systems where traditional chemical reactors or fermentors are integrated with one or more membrane separation units (RO, UF, PV, etc.) will be evaluated. Depending on the immobilization type, membrane configuration, and fluid dynamics, the following main catalytic membrane reactor configurations might be considered:

- Enzyme membrane reactors and fermentors, where enzymes are continuously flushed along membranes

- Catalytic membrane reactors with segregated enzyme, where the enzyme is segregated within the membrane module (in the lumen or in the shell side) or entrapped within the membrane pores

- Catalytic membrane reactors with gelified enzyme, where enzyme is immobilized in a proteic gel layer, dynamically or statically formed

- Enzyme-bound membrane reactors, where the enzyme is chemically bound to the membrane surface or in the pores.

Membrane reactors are also named on the basis of type of solvents and/or separation process utilized. For example, 'biphasic organic/aqueous membrane reactors', where the enzyme is physically or chemically immobilized on the membrane surface or within the pores, and two immiscible phases flow along the membrane (see Chapter 6); or 'affinity membrane reactors', where a chemical reaction (carried out with any of the previously listed membrane reactors) is coupled to a separation by affinity membrane (Klein, 1990). (Examples of this type of reactor are reported in Chapter 4, where the 'affinity membranes' are also enantioselective.) Various drawbacks that characterize the traditional enzyme immobilization techniques, such as reduction of initial enzyme activity due to the binding process, diffusional resistances, difficulties in coenzyme-dependent reactions, etc., can be minimized using the various enzyme membrane reactor configurations.

2.2 Enzyme Membrane Reactors and Fermentors

A definition of an enzyme membrane reactor (EMR) might be a reactor system in which an appropriate membrane separation is used to keep larger components (i.e. enzymes and/or macromolecular substrates) in the reactor vessel while low molecular mass molecules (i.e. products and/or inhibitors) are allowed to pass freely through the membrane, thus leaving the reactor as permeate. In this set-up, several advantages of immobilized preparations, together with easy recovery of deactivated enzymes and replacement with fresh catalysts, are achieved; moreover, inhibitors are continuously removed from the reaction vessel. The direct and deep contact between substrates and biocatalysts limits diffusional resistances, while no activity losses due to fixation to the support occur, thus maximizing the activity of the biocatalyst. On the other hand, enzyme stability is not improved.

Both dead-end and continuous stirred tank reactor (CSTR) UF cells with flat membranes have been proposed as enzymatic reactors. In Figure 2.3 the set-up of a continuous stirred tank membrane reactor is presented.

The performance of dead-end units is largely affected by the flow dynamics of the system; in fact, mixing of substrates and catalysts is not fully accomplished, and concentration polarization phenomena strongly limit reactor performance.

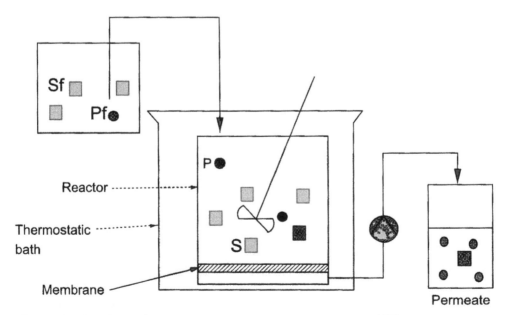

Figure 2.3 Experimental set-up of an enzyme membrane reactor. (Sf(■) = substrate feed concentration; Pf(●) = product feed concentration; S(■) = substrate concentration; P(●) = product concentration.)

Continuous stirred tank reactors have been more widely adopted, partly owing to the possibility of control of concentration polarization and partly owing to the easy modelling of enzyme kinetic behaviour. In the literature, comprehensive mathematical descriptions of the kinetic behaviour of enzymes located in CSTR UF units are reported (Wichmann et al., 1981; Hagerdal et al., 1980).

Assuming complete mixing within the reactor so that enzyme and substrate concentrations in the reactor vessel are uniform, and the latter is equal to its value in the permeate, substrate mass balance in molar form can be written as

$$\frac{S_f - S}{\tau} = R(S,P) \tag{2.1}$$

where

S_f = feed substrate concentration $[ML^{-3}]$
S = substrate concentration $[ML^{-3}]$
τ = reactor time constant $[T]$
R = reaction rate $[MT^{-1}L^{-3}]$

Product steady-state mass balance will similarly be

$$\frac{P - P_f}{\tau} = R(S,P) \tag{2.2}$$

where

P = product concentration $[ML^{-3}]$
P_f = feed product concentration $[ML^{-3}]$

Assuming that substrate conversion obeys the simple Michaelis–Menten model, substrate steady-state mass balance reduces to

$$XS_f + \left(\frac{X}{1 - X}\right)K'_M = KEV/Q \tag{2.3}$$

with R being expressed as $KES/(K'_M + S)$ and where

X = $(S_f - S)/S_f$ = degree of conversion
K'_M = enzyme Michaelis–Menten constant $[ML^{-3}]$
K = enzyme kinetic constant $[T^{-1}]$
E = enzyme concentration $[ML^{-3}]$
V = reaction volume $[L^3]$
Q = flow rate $[L^3T^{-1}]$

The equation in this form is a useful tool for estimating parameters of reaction kinetics. Instead of performing nonlinear parameter estimation procedures, the functional dependence of XS_f on $X/(1 - X)$ can be plotted. The plot should be a straight line whose slope is $(-K'_M)$ and whose intersection with the XS_f axis gives the coordinate of $V_{max}\tau$. Hence, it furnishes a rapid graphic procedure for obtaining rough estimates of kinetic parameters.

When more complex kinetics are involved, so that the substrate rate of conversion is dependent on both substrate and product concentrations, combining equations (2.1) and (2.2) with the rate equation eventually leads to an implicit equation (Hong et al., 1981):

$$\frac{P - P_f}{\tau} = R(P) \tag{2.4}$$

Product concentration in the permeate can then be determined either with numerical procedures or using graphic techniques. True reactor operating point at steady state is given by the intersection point of the straight line representing mass balance and the curve of the reaction term. If multienzymatic systems are used as the catalyst, it must be considered that the reaction term is the course of the reaction rate relative to the substrate conversion of the key component.

Enzyme activity is usually not constant with time. Physicochemical changes in enzyme structure, thermal denaturation and microbial contamination cause enzyme activity to decrease continuously with time. When enzymes or cells are compartmentalized in UF cells, biocatalyst losses can even occur owing to the wrong choice of membrane molecular mass cut-off. It is conventional to measure the enzyme stability in terms of its half-life time ($t_{1/2}$), that is the time in which enzyme activity is reduced to half its initial value. It can be calculated from the following equations:

$$K_d = \frac{2.303}{\theta} \log \frac{A_{E_0}}{A_{E_\theta}}$$

$$t_{1/2} = \frac{0.693}{K_d}$$

where

K_d = enzyme deactivation constant $[T^{-1}]$

θ = operation time $[T]$

A_{E_0} = initial enzyme activity, or product mass per unit time $[MT^{-1}]$

A_{E_θ} = enzyme activity at time θ $[MT^{-1}]$

Since biotransformations by means of enzymes are continuous processes, as long as the reactor working life is longer than the native enzyme half-life, enzyme activity decay with time must be taken into account in order to assess correctly reactor performance. A transient substrate mass balance on the CST reactor leads to

$$V \frac{dS(t)}{dt} = Q[S_f - S(t)] - V_{max}V \tag{2.5}$$

where

V = reaction volume $[L^3]$

S = substrate concentration $[ML^{-3}]$

S_f = feed substrate concentration $[ML^{-3}]$

V_{max} = enzyme reaction rate at saturating substrate concentration $[ML^{-3}T^{-1}]$

Q = flow rate $[L^3T^{-1}]$

Let us assume that substrate inlet concentration is much higher than the enzyme's apparent Michaelis constant, K_M', so that enzyme kinetics are expressed according to

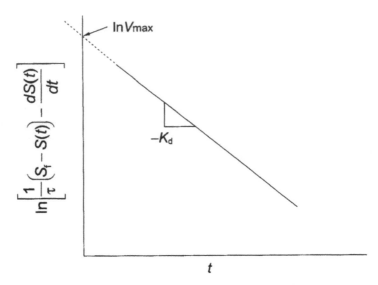

Figure 2.4   Determination of V_{max} and K_d values (from equation 2.7).

zero-order kinetics, i.e. $R = KE = V_{max}$. Enzyme activity decay can be expressed in terms of the Arrhenius equation:

$$E = E_0 \exp(-K_d t) \tag{2.6}$$

where

E = enzyme concentration $[ML^{-3}]$
E_0 = initial enzyme concentration $[ML^{-3}]$
K_d = enzyme deactivation constant $[T^{-1}]$

Equation (2.5) then takes the form

$$\frac{S_f - S(t)}{\tau} - \frac{dS(t)}{dt} = V_{max_0} \exp(-K_d t) \tag{2.7}$$

Integration of the differential equation (2.7) leads to an estimate of the outlet substrate concentration:

$$S(t) = \frac{V_{max_0}[\exp(-t/\tau) - \exp(-K_d t)]}{(1/\tau - K_d) + S_f} \tag{2.8}$$

Results presented in the literature (Wisniewski et al., 1983) show good agreement between the theoretical prediction and the experimental data.

An estimate of the deactivation constant can be made graphically from equation (2.7) by plotting $\ln\{[S_f - S(t)]/\tau - dS(t)/dt\}$ vs t. A straight line is thus obtained whose slope is given by $(-K_d)$ and the line intersects the vertical axis at the point of the coordinate of $\ln V_{max}$ (Figure 2.4).

When the reacting solution is fed into the system, low molecular mass products leave the reactor, permeating the membrane, whereas enzymes are partially or totally rejected. Then, enzymes tend to accumulate in a thin layer immediately upstream from the membrane, causing polarization phenomena. The extent to which concentration polarization affects reactor performance depends on the balance of rejected solutes (e.g. enzymes or cells), accumulation due to membrane rejection and back-diffusion to the bulk phase, and eventually on the flow dynamics of the reacting vessel.

For macromolecules (like most biocatalysts), if membrane properties are carefully chosen, membrane rejection is usually very good, while biocatalyst back-diffusion towards the bulk phase is extremely slow. The effect of concentration polarization on such reactor performances can then be significant (Blatt et al., 1970).

Applying the thin-film theory to a region immediately upstream from the membrane results in the following steady-state mass balance equation:

$$JE = -D_e \frac{dE}{dx} \tag{2.9}$$

under the boundary condition $x \to \infty$, $E = E_s$, and where

J = volumetric flux $[L^3 L^{-2} T^{-1}]$
E = enzyme concentration $[ML^{-3}]$
E_s = enzyme concentration in the bulk liquid phase $[ML^{-3}]$
D_e = enzyme diffusion coefficient $[L^2 T^{-1}]$

Under the assumption of total rejection of enzyme macromolecules, integration of equation (2.9) allows one to express the ratio of enzyme concentration in the bulk solution in the presence of concentration polarization to its value in the absence of concentration polarization phenomena as

$$\frac{E_S}{E_0} = \{(1 - \alpha) + (\alpha/\beta)[\exp(\beta) - 1]\}^{-1} \tag{2.10}$$

Enzyme concentration at the membrane–solution interface can thus be related to enzyme concentration in the bulk phase by

$$E_w = E_s \exp(\beta) \tag{2.11}$$

where the dimensionless parameter β is a measure of convective mass transfer, J, relative to the overall mass transfer, D/δ, more generally K_s (Mathiason and Sivik, 1980).

The change of enzyme bulk concentration can be dramatic with changes in permeate flow rate and applied pressure (Hong et al., 1981). Changes in enzyme bulk concentration may induce a large reduction in reaction rate within the membrane reactor. Under given stirring conditions, therefore, a critical value of flow rate and hence of applied pressure exists. At flow rates lower than the critical value, the reactor always attains steady operation conditions; and correspondingly outlet product concentration is constant. Beyond the critical value, concentration polarization phenomena promote the localization of a large fraction of enzymes near the membrane surface, seldom leading to enzymatic gel formation. Correspondingly, an accelerated deactivation of enzyme activity is superimposed on reactor performance, thus hindering the attainment of steady-state conditions. Experimental evidence suggests that the contribution of polarized enzymes to overall conversion can be negligible, owing to the consistency of product concentration in the bulk phase of the reactor and in the permeate.

When macromolecular substrates are involved in the transformation under study, concentration polarization phenomena affect EMR performance more severely. Diffusion limitations of macromolecular substrates hamper the use of immobilized enzymes in the hydrolysis of high molecular mass substrates. By selecting membranes with an appropriate molecular mass cut-off, both enzyme and substrate are retained in an EMR in contact with each other, and hydrolysis products and/or inhibitors are continuously removed from the system. Soluble enzymes can then act directly on substrate macromolecules without diffusion limitations and steric hindrance imposed by enzyme fixation to a solid support. The stirred characteristic of CST EMRs moreover ensures that substrates and/or inhibitors within the reactor vessel are maintained at the lowest possible concentration level. Such a reactor configuration is extremely useful when substrate-inhibited reaction patterns are involved, or when inhibiting species are assumed to exist in the feed stream.

Diafiltration, semicontinuous and continuous operational modes have been proposed (Ohlson et al., 1984). Operating the reactor in a semi-continuous condition and adding substrate so as to keep its bulk concentration constant leads to significant changes in reactor performance compared to the diafiltration mode. Under both operational modes, permeate flow rate continuously decreases with time.

When EMRs are operated continuously, feeding a slurry of substrate macromolecules to the reactor, concentration polarization phenomena play a dominant role. The presence within the reaction vessel of contaminants or intermediate products which are not fully hydrolyzed by the enzymatic system under study can lead to membrane fouling or to the formation of a gel layer at the membrane surface. Under such conditions, the filtration rate decreases continuously with time, and it may happen that substrate conversion does not attain steady-state conditions. The addition of enzymes capable of hydrolyzing such foulants to low molecular mass compounds usually improves reactor performance, eventually approaching steady-state conditions in terms of both permeate flow rate and substrate conversion. In addition, antifouling procedures including suitable feed pretreatment or procedures to reduce concentration polarization can be used (see Chapter 1).

Equation (2.11) demonstrates the dependence of enzyme concentration at the upstream membrane surface on the flow dynamics of the reaction vessel. Owing to the monotonicity of the exponential function, at given applied pressure operational conditions that improve mass transfer in the bulk phase (i.e. increasing axial flow rate improves K_S or D/δ) diminish concentration polarization. High stirring speeds usually improve filtering performance of a CST flat membrane reactor, albeit at the expense of partial enzyme deactivation.

Reactor configurations other than dead-end or CSTR can offer improvements.

The use of hollow-fibre ultrafiltration modules in crossflow filtration mode strongly improves both reactor performance and economics. Capillary membranes are characterized by a favourable surface-to-volume ratio; advantages from this feature relate not only to lower overall plant size but also to the increase of the surface-to-price ratio. Flushing of enzyme/substrate solution through a UF module tangentially to the membrane surface at a high linear velocity reduces the extent of concentration polarization, avoiding the formation of a secondary enzymatic gel membrane, in spite of product flux through the membrane (Figure 2.5). Provided that the volume of the UF module is small relative to the total volume, and that recirculation flow rate is much larger than permeate flux, system kinetic behaviour can be modelled in terms of a CST reactor. Reactor productivity as a function of time can be related to flux and enzyme concentration according to the following equation (Deeslie and Cheryan, 1981):

$$P_r = \frac{X(t)S_f J(t)}{EV} \tag{2.12}$$

Reactor productivity (g product/g enzyme) is affected by flux and enzyme concentration. For example, for soy protein hydrolysis performed by Pronase enzyme, one maximum of productivity can be obtained by operating the reactor at the highest flux and the lowest enzyme concentration.

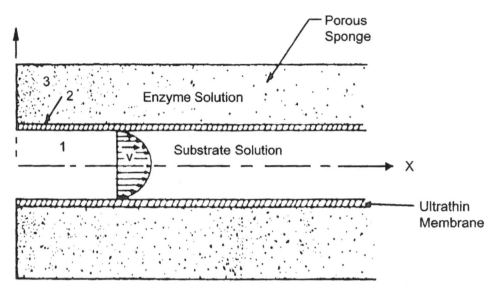

Figure 2.5 Axial section of an asymmetric hollow fibre under crossflow filtration mode.

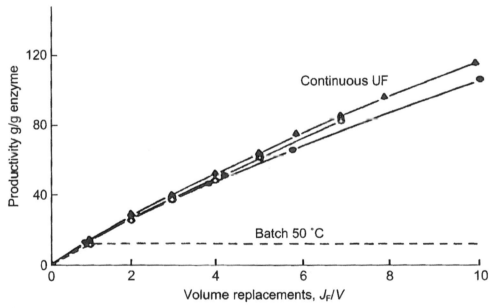

Figure 2.6 Comparison between productivity of a continuous UF reactor and a batch reactor for the Promine D-Pronase system at pH 8.0 (Cheryan and Mehaia, 1983).

Comparisons between performance of batch reactors (Cheryan and Mehaia, 1983) and continuous membrane reactors (with hollow fibres) are reported in Figure 2.6. In the continuous process, enzymes are charged at the beginning of the process, and the system is kept working until enzyme activity drops to a given limit. In batch reactors, the enzymes must be replaced at the end of each cycle. Hence the longer the reactor is operated, the greater is the productivity and the larger the difference between continuous and batch system performance. Moreover, batch processes require added expenses for enzyme inactivation and product purification, as well as more labour.

Patents already exist, for example for amino acid production from α-ketoacids, which confirm the potential for scale-up of EMR plants (Wandrey et al., 1982).

Enzyme membrane reactors in which enzymes are flushed along the membrane are preferred over other immobilized preparations when homogeneous catalysis with multi-enzyme systems is required. This reactor configuration has the advantages of no diffusion limitations; absence of enzyme fixation costs; sterilizability of the plant; and constant productivity assured by enzyme dosage.

The overall benefits of EMRs might, however, be reduced because the control of concentration polarization phenomena on an enzyme membrane system might be less efficient than necessary. Furthermore, in a turbulence regime, in general, high axial flow rate can induce removal of the catalyst, its denaturation by shear effects, and reduction of residence time of substrate to the catalyst native site. Optimization of the fluid dynamic regime in CMR is therefore necessary.

Thus far, EMRs have been used successfully with macromolecular substrates, as for the saccharification of cellulose (Ohlson et al., 1984) and protein hydrolysis (Deeslie and Cheryan, 1981), or with low molecular mass substrates, as for L-malic acid production from fumaric acid (Leuchtenberger et al., 1984).

2.2.1 Membrane Fermentors

A particular case of membrane reactors with the biocatalyst flushed along the membrane are the membrane fermentors, where microporous membranes are used to separate the fermentation broth from the product stream, thus retaining viable cells in the fermentor.

A typical apparatus is shown in Figure 2.7 in which there is a vessel for the growing biomass, where the pH and temperature are strictly controlled and nutrients are added, and a UF unit, with membranes in hollow-fibre or flat-slab configuration, to withdraw products from the flowing cell slurry.

When the biomass is well developed, the reactor biomass is pumped to the UF unit where solid–liquid separation occurs. The sludge is flushed back to the reactor. In most cases, the flow rate of nutrient feed is kept equal to the permeate flow rate, thus maintaining

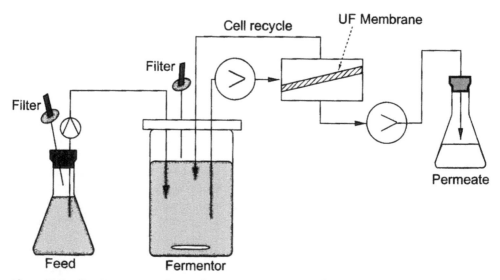

Figure 2.7 Continuous membrane fermentor apparatus with cell recycle.

a constant liquid level in the reactor. In this case, the continuous membrane fermentor is also referred to as a cell recycle membrane fermentor (which differs from the case where the cells are compartmentalized within the membrane module or within the membrane itself; Figure 2.9, see below).

As described for enzymes, viable cells can also be confined within the shell of a hollow-fibre membrane module or in the pores of microfiltration membranes. When kinetic models for the growth and fermentation of a specific kind of cell or microbe on the corresponding substrate medium are available, mathematical modelling of membrane fermentors can be accomplished according to mass balances and the equations reported for the corresponding enzyme reactor configurations examined in the previous sections.

A steady-state mass balance on a continuous cell recycle fermentor over the vessel alone (see Figure 2.7) is

$$\alpha'FX'_2 - (1 + \alpha')FX'_1 + V\mu^*X'_1 = 0 \tag{2.13}$$

on cells, and

$$X' = \frac{YD'(S_f - S)}{\mu^*} \tag{2.14}$$

on limiting growth substrate to give a cell concentration value in the vessel equal to

$$X'_1 = \frac{Y(S_f - S)}{(1 + \alpha' - C\alpha')} \tag{2.15}$$

where

α' = recycle ratio
F = feed flow rate $[L^3 T^{-1}]$
X' = cell concentration $[ML^{-3}]$
X'_1 = cell concentration entering the membrane module $[ML^{-3}]$
X'_2 = cell concentration in the recycle stream $[ML^{-3}]$
μ^* = specific cell growth rate $[T^{-1}]$
Y = organism yield coefficient, i.e. mass of cells formed/mass of nutrients
C = concentration factor

The specific growth rate (SGR) can be expressed as a linear function of the specific substrate utilization rate, KS, to give (Li and Sutton, 1982)

$$SGR = Y \cdot KS - D \tag{2.16}$$

where

KS = the specific substrate utilization rate [mass/mass time]
D = the organism decay coefficient [1/time]

The kinetics of substrate removal in the anaerobic reactor actually determines the solid retention time (SRT) required for a given fermentation efficiency. In turn, the presence of the UF unit allows fine control of SRT, making process control easier. When high biomass concentrations are achieved, longer SRTs can be maintained at lower reactor volumes.

The volumetric loading to a reactor, VL, is defined as (Reach et al., 1982)

$$VL = \frac{QS_f}{V} = \frac{S_f}{\tau} \tag{2.17}$$

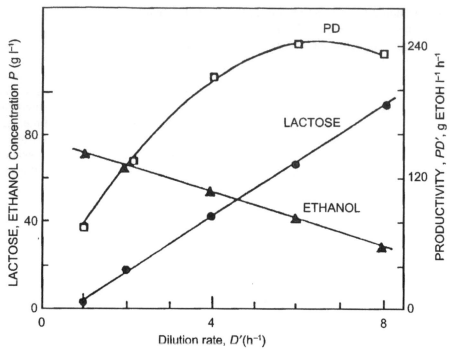

Figure 2.8   Fermentation kinetics of a membrane recycle fermentor.

High biomass concentrations permit operation at higher volumetric loading rates. Reactor design can be carried out by estimating the value of volumetric loading necessary to achieve a given effluent quality. This design criterion is commonly used for biological systems where the evaluation of biomass concentration within the reactor is difficult. This is not the case with cell recycle fermentors, so the VL or the SRT approach can also be used as a design criterion.

Figure 2.8 shows the typical dependence, at steady state, of substrate concentration, product concentration and productivity as a function of dilution rate in a cell recycle membrane fermentor. As dilution rate increases, fermentor productivity increases, attains a maximum value and then decreases. In the permeate, product concentration steadily decreases and substrate concentration increases as dilution rate increases. A compromise generally has to be made between production rate and product concentration in the effluent. Low substrate concentration and high product concentration in the permeate result in low recovery costs of products from the effluent stream. The optimum operating conditions have to be determined by the economics of the overall process.

In continuous membrane fermentors productivity is a function of viable cell concentration within the fermentor. Cell recycle allows operation at higher microbial cell concentrations than in conventional batch fermentors, reducing the volume required for a given productivity and hence the capital costs. On the other hand, the viscosity of the cell slurry increases with cell concentration, and concentration polarization phenomena are usually significant.

Systems such as rotating membrane modules, back-flushing operations, etc., can be used to overcome fouling and expected concentration polarization problems. Provided cell resistance to shear stress is good, concentration polarization is usually controlled by operating the reaction system at high recirculation rates.

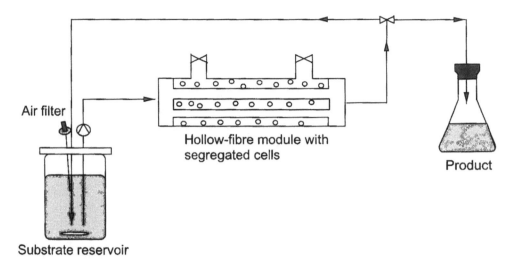

Substrate reservoir

Figure 2.9 Schematic of a continuous membrane fermentor with segregated cells in a hollow-fibre fermentor.

When concentration polarization occurs, permeate fluxes become invariant with the transmembrane pressure, and increases in the permeation rate can be achieved only by appropriate fluid dynamic conditions. The design of the UF membrane unit in the polarized regime must relate the flux to the optimal level of reactor volatile suspended solids (VSS) or total suspended solids (TSS) according to Michaels' gel theory, to give

$$J = K_S \ln(C_g/C_b) \tag{2.18}$$

where

K_S = the overall mass transfer coefficient $[LT^{-1}]$

C_g = apparent solid concentration in the gel $[ML^{-3}]$

G_b = concentration in the bulk, in the specific case, concentration of VSS or TSS $[ML^{-3}]$

High recirculation and dilution rates help in maintaining low levels of inhibitory products.

Continuous fermentation processes can also be carried out in a hollow-fibre fermentor (HFF) with cells segregated within the shell side, while substrate solution feeds to the core of the fibres (Figure 2.9) (these membrane reactor systems will be described in detail in the next section, where enzymes are taken as biocatalyst models). Since the cell slurry is separated from the substrate solution by the membrane, HFFs can be used for fermentation of liquid streams containing low molecular mass fermentable substances. A careful choice of membrane molecular mass cut-off can also ensure that the cell environment is fully sterilized. The behaviour of the fermentor is similar to that of a cell recycle reactor with the same fermentation system. Higher productivity and yields than with batch fermentation are obtained. The long-term cell stability in a HFF is also better than in a batch fermentor. However, even with HFF, reactor productivity is increased at the expense of low substrate conversions, at least at low dilution rates.

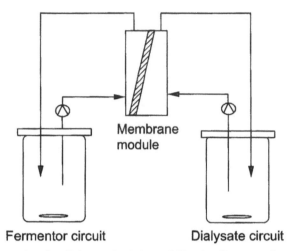

Figure 2.10 Schematic of a dialysate-feed, immobilized cell system for dialysis continuous fermentation.

Table 2.1 Performance of different membrane fermentor configurations

Substrate	Reactor	Productivity $(g\ l^{-1}\ h^{-1})$	Product concentration $(g\ l^{-1})$	Reference
Lactose	Batch	25	15	Cheryan and Mehaia (1983)
Lactose	HFF	63	15–25	Mehaia and Cheryan (1984)
Whey lactose	HFF	35	5–10	Mehaia and Cheryan (1984)
Whey lactose	HFF	50	37.5	Timmer and Kromkamp (1994)
Glucose	Batch	1–2	45	Vick Roy et al. (1983)
Glucose	CSTR + HF	76	35	Vick Roy et al. (1983)
Glucose	HF reactor	100	2	Vick Roy et al. (1983)

In another membrane fermentor configuration, the cell slurry is separated from substrate solution by a dialysis membrane (Figure 2.10). Two different operational modes for this configuration have been proposed. In the first, substrate is fed into a continuous fermentor circuit that is dialyzed against a continuous dialysate circuit where water is kept flowing. In the second, substrate is fed into the dialysate circuit and diffuses towards the batch fermentor circuit through a dialysis membrane.

The performance of different membrane fermentors is reported in Table 2.1.

A new membrane fermentor design has also been suggested in which the microporous membranes have been directly incorporated inside the fermentor to allow the extraction *in situ* of the fermentation products able to induce inhibitory phenomena (Figure 2.11). Microporous hydrophobic hollow fibres have been incorporated along the length of a tubular fermentor in ethanol production.

This system is particularly useful in aerobic fermentations, where the membrane module can also be used to supply air to the microorganisms, since reduced fouling problems are also obtained (Qian Yi, personal communication, 1997; Strachan and Livingston, 1997).

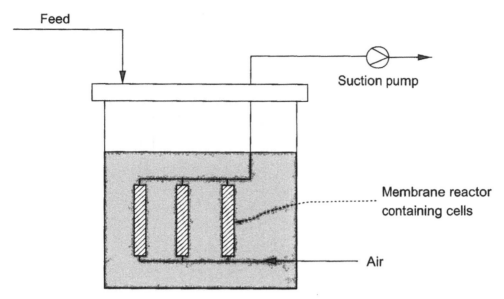

Figure 2.11 Membrane fermentor with microporous membranes directly incorporated inside the fermentor.

2.3 CMRs with Segregated Enzyme

In some cases membrane reactors can be operated without a crossflow regime and the enzyme or cells are therefore compartmentalized in the shell of the modules or in the lumen of the fibres. In these reactors, enzymes or cells are not linked but only confined in a defined region of membrane module space. The segregation of biocatalyst is achieved by means of membranes with a suitable molecular mass cut-off. In this way, enzymes and bacterial cells are not lost in the effluent stream, and low molecular mass products and inhibitors can be removed through the membrane (see Figure 2.9).

The development of hollow fibres with diameters down to about 100 µm makes possible tube-and-shell reactors with high surface-to-volume ratio. Biocatalytic reactors can segregate enzymes or cells either within the hollow-fibre lumen, within the shell surrounding the outer surface of the fibres, or within the porous membrane support.

Segregated enzyme reactors avoid the negative aspects of immobilization techniques such as steric hindrance and enzyme deactivation due to coupling or shear stresses.

There is also growing interest in therapeutic applications of compartmentalized cells or microsomes functioning as a bioartificial pancreas or an extracorporeal detoxification device (Goosen et al., 1985; Takabakate et al., 1991).

A theoretical analysis of such enzyme membrane reactors was carried out by Rony (1971) and Waterland et al. (1974) for the case where asymmetric membranes were used and the biocatalyst was confined in the lumen (Figure 2.12a) and is also discussed by Drioli et al. (1989). Four different regions were distiguished in the reacting system (Figure 2.12b): the lumen of the fibre (1) where enzyme solution is retained; the dense thick layer (2); the porous spongy part of the membrane (3); the shell region where substrate solution flows (4).

Evaluation of the stability and catalytic properties of the immobilized system must take into account possible pH differences between the inner core of the fibre, where the

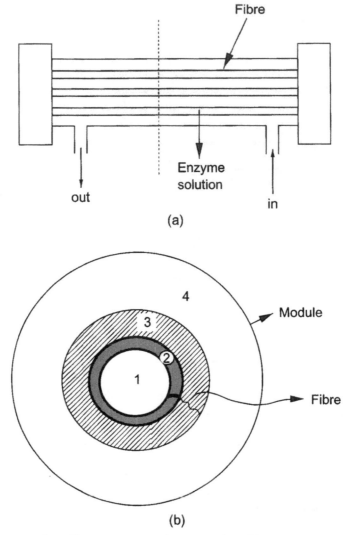

Figure 2.12 (a) Hollow-fibre enzyme membrane reactor with enzyme segregated within the lumen. (b) Cross-section of hollow-fibre enzyme compartments. 1 = lumen; 2 = dense layer; 3 = sponge layer; 4 = shell.

reaction takes place, and the bulk of the feed solution. Experimentally, these differences produce more or less pronounced shifts in the optimum pH dependence of enzyme activity relative to its free form dependence and thereby affect the activity of the enzyme at work.

Similar reactor configurations using flat membranes in place of hollow fibres have been used with urease, uricase, glucose oxidase and creatinine kinase.

An enzyme solution is, in this case, introduced into one of the membrane-separated chambers of a flat-membrane dialyzer. Operating the reactor at high feed flow rates leads to a reduction of mass transfer resistances, which can also be achieved using suitable turbulence promoters. As with enzymes, whole active microorganisms can be segregated in a definite region of space by means of membranes in order to catalyze specific reactions. Microsomes and bacteria have been and are currently employed in membrane reactors to perform complex multienzymatic reactions or to reduce overall reactor costs, avoiding enzyme purification. When the size of the biocatalyst exceeds the dimensions of the pores in the sponge of the asymmetric membranes, the biocatalyst can be compartmentalized to stay on the shell side (or lumen side), while the substrate solution is kept flowing within the lumen (or shell) of the fibres.

Apparently the main transport mechanisms through which substrate conversion takes place are diffusion of substrates from the bulk fluid phase to the membrane; diffusion of substrates within membrane pores; and diffusion of substrates within the shell-side of the reactor to the biocatalyst.

Referring to a single fibre, the scheme of the reacting system is similar to those examined so far. However, the reaction does not occur in either regions 1, 2 or 3 (Figure 2.12b). The differential mass transfer and continuity equations defining substrate and product concentration in these regions are equivalent to those previously examined. The description of mass transport in the shell-side region is somewhat more complicated. Differences in the environment surrounding each fibre, the position of fibres in the bundle, and the ultrafiltration fluxes make both the analytical and the numerical approaches quite difficult.

Models proposed often deal with simpler systems, assuming good mixing conditions both in the shell and within the lumen of the fibre, and pure diffusive fluxes. In some particular configurations, as in the artificial pancreas, such assumptions hold and the models work quite well, especially in the case of membrane units using one large hollow fibre alone. In cases where a quick, transient response is needed, the only way to circumvent the slow transient behaviour of the device is to reduce the volume wherein the catalyst is compartmentalized. Even though diffusion appears to play an important role in substrate and product transport, there is experimental evidence that the bundles of hollow fibres assembled in a 'tube-and-shell' configuration respond more quickly than would be predicted assuming purely diffusive fluxes across the membrane walls (Breslau and Kelcullen, 1975). Pressure drop along the length of each fibre should therefore produce a transmembrane pressure across the membrane wall such that, at the inlet of the reactor, the pressure difference promotes an ultrafiltration flux towards the shell side, where the catalyst is, and at the outlet it promotes a backward flux from the shell towards the lumen of the fibre (there is only a small pressure drop in the shell). These pressure profiles promote fluxes that improve the performance of the system compared to that exhibited by pure diffusive reactors.

A quantitative analysis of such ultrafiltration flux has been approached in the case of a single fibre device (Reach et al., 1982).

This reactor configuration is often appropriate for complex catalytic systems. The use of cofactors with the enzymes in continuous flow systems has often been limited by the need to supply large amounts of fresh cofactor, usually an expensive compound. Many procedures have been suggested for confining these low molecular mass compounds in a well-defined region of space where they are continuously used and regenerated. When the biocatalyst is compartmentalized in this way, cofactor costs are reduced. In the presence of a suitable regeneration system, only low cofactor additions are needed to maintain excess concentration levels, thus ensuring maximum rates of conversion (Chambers et al., 1976).

Reactors in this configuration are also employed in therapeutic applications. Bundles of hollow fibres or a single large hollow fibre in a cylindrical module are used to separate blood flowing within the lumen and mammalian pancreatic islets as assistance to diabetic patients. They have also been suggested for use as extracorporeal blood detoxifiers, where mammalian liver microsomes are compartmentalized in the shell of the module. In such systems, if the membrane molecular cut-off is chosen carefully, membranes will protect transplanted cells from the immunodefensive action of leukocytes in the blood.

The main drawback of membrane reactors in this configuration is the relatively slow response to metabolic stimulation due to the large shell volumes required to accommodate the number of large biocatalytic units needed to offer the required assistance.

Diffusion and ultrafiltration fluxes due to pressure drop along the length of fibres play the most important role in substrate and product mass transfer when systems are operated as previously described. However, an ultrafiltration flux can be promoted from the lumen of the fibres outwards and/or from the shell inwards. Better reactor performances should result from such operating conditions.

2.3.1   CMRs with the Biocatalysts within the Pores of Asymmetric Membranes

Membranes have been cast with microbial cells incorporated in the casting solution (Drioli et al., 1982). To date, the casting of UF and RO membranes charged with enzymes or whole cells on an industrial scale has been limited by the drastic conditions under which the synthetic membranes are usually formed. The presence of nonaqueous solvents in casting solutions, and the high temperatures required by membrane annealing, generally denature enzymatic proteins, with a loss in their catalytic activity. Recently, a number of microorganisms, e.g. *Sulfolobus solfataricus*, have been discovered which can withstand both high temperatures and organic solvents. Microbial enzymes maintain their activity in conditions that are otherwise denaturating, presumably because of cellular membrane protection. Polysulfone and cellulose acetate membranes have been cast with microbial cells in the casting solution using the phase-inversion technique, as well as in polyurethane foams. Cell-loaded membranes appear to be kinetically active and stable over long periods. It is noteworthy that cell entrapment can enhance microbial activity compared to cell behaviour in homogeneous solution, an effect probably due to cellular membrane permeabilization as a consequence of the entrapment procedure.

Entrapped whole cells are the source of a number of microbial enzymes useful for industrial purposes. In addition, the possibility of long-term operation make this immobilization procedure extremely attractive.

Asymmetric synthetic hollow-fibre membranes designed for use in ultrafiltration/dialysis processes can provide interesting supports for immobilizing enzymes. The use of these fibres in bioreactors can eliminate some of the disadvantages of the reactor systems discussed previously.

Enzymes can be entrapped within the outer sponge layer of the fibres by crossflow filtration of an enzyme solution. If the pores in the dense layer are small enough to retain enzyme molecules but large enough to pass substrates and products freely, the enzymes are effectively immobilized or segregated within the spongy annular section (Figure 2.13). The amount of immobilized protein can be determined by mass balance (Giorno et al., 1995):

$$C_i V_i = C_p V_p + C_r V_r + m \tag{2.19}$$

where

$C_i V_i$ = initial mass [M]
$C_p V_p$ = mass in the permeate [M]
$C_r V_r$ = mass in the retentate [M]
m = mass in the membrane [M]

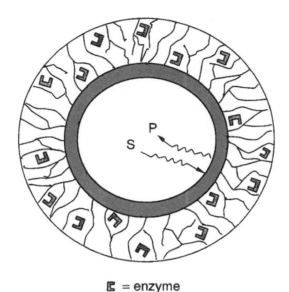

$\mathbb{C}$ = enzyme

Figure 2.13 Schematic of the cross-section of a hollow fibre with enzyme immobilized in the macropores.

Once the enzyme is deactivated, it can be removed by back-flushing ultrafiltration of sodium hydroxide solution.

The reactor can be operated to ultrafilter the substrate solution from shell to lumen, without applying a transmembrane pressure in this direction. In this case, the dynamics of substrate conversion depend on enzyme kinetics as well as on mass transport conditions. Diffusion through the membrane matrix and within the flowing solution plays the most important role in transport mechanisms. Since the flow is laminar in most cases, substrate and product transport resistances through the dense layer are exceedingly small relative to diffusional resistances in the flowing solution. The rate-limiting step in substrate conversion is therefore either simple diffusion or the intrinsic kinetics of the reaction itself.

Mathematical modelling of such reactors has been investigated extensively (Waterland et al., 1974; Lewis and Middleman, 1974; Davis, 1974; Davis et al., 1974). Lewis and Davis have proposed an analytical solution to the problem in the case of low feed substrate concentration, that is for a linear rate equation. An iterative numerical solution for the nonlinear problem was fully developed by Waterland et al. (1974).

An increase in feed substrate concentration may shift reactor operation from a mainly diffusion-controlled regime to a mainly kinetically controlled regime, with a consequent decrease in reactor conversion.

Experimental work in which enzymes with relatively simple kinetics, e.g. α-galactosidase, invertase, glucose isomerase and urease, were immobilized in the sponge of hollow fibres are in good agreement with predictions of the theoretical model. Quick graphical procedures are available in the literature for evaluating the extent to which external and internal diffusion affect immobilized enzyme kinetics (Engasser, 1978; Bailey and Ollis, 1986). Agreement between experimental results and predictions of the more comprehensive proposed model suggests that it might be used to predict reactor performance in the case of simple kinetics and when the kinetic and transport parameters are known.

When complex kinetics are involved in substrate conversion, as with product-inhibited enzymes (such as amyloglucosidase, which catalyzes maltose conversion to glucose) or with reactions involving a number of intermediates (such as starch hydrolysis by means of amyloglucosidase), definitive information on enzyme kinetics is rarely available. Moreover, when solutions of macromolecular compounds, like starch, are fed to the reactor, diffusional resistances are more pronounced and may hinder the possibility of using such reactors for analytical purposes. A thorough knowledge of the chemical mechanisms through which substrates are converted to products and of the coupling of transport phenomena to enzyme reactions appears to be a prerequisite for the design of such biochemical reactors.

Different operating conditions may require some modification of the analysis. Ultrafiltration and/or osmosis can promote convective solute or water flux through the membrane wall. Should this happen, radial convection could compete with diffusion as the main substrate and product transport mechanism. The relative importance of the two transport mechanisms can be evaluated by comparing the radial convective velocity to the diffusive velocity, that is the ratio of the diffusion coefficient to the wall thickness. When the former is negligible relative to the latter, the model applies without modification. The second possible effect of the radial flux is to remove enzymes from the fibre wall, resulting in the reduction of reactor efficiency.

This reactor configuration is particularly attractive since substrates are physically separated from the enzyme solution only by a very thin membrane layer, the dense skin, thus minimizing mass transfer diffusional resistances. A number of other advantages make this reactor configuration feasible for many applications. The use of small-diameter fibres leads to large area-to-volume ratios, with high enzyme loading capacity per unit of reactor volume. The enzyme microenvironment is fully shear free. Moreover, membranes act as selective barriers, protecting enzymes from macromolecular contaminants such as proteolytic enzymes or selecting substrates on the basis of their permeability or electric charge.

Many enzymes can even be co-immobilized with the macroporous region of asymmetric membranes. The procedure itself does not exclude the possibility of cross-linking enzymes directly to the porous polymeric matrix of the fibres, nor that of compartmentalizing within the same region enzymes previously coupled to soluble polymers or inactive proteins. Such binding procedures might in fact result in a greater stability, that is a slower decay of enzyme activity with time.

In this reactor configuration, once products are formed they diffuse back towards the stream flowing in the core of the fibres. Such configuration is therefore feasible for all those applications in which products at relatively high concentration are tolerated in the circulating stream.

2.4 CMRs with Gelified Enzyme

Ultrafiltration of protein solutions is a proven unit operation for obtaining purified enzymes from cell cultures. Under certain circumstances, such processes can provide an interesting technique for enzyme immobilization that takes advantage of concentration polarization phenomena. If an enzyme solution is flushed under pressure through an ultrafiltration membrane that completely rejects enzyme molecules, these molecules will accumulate on the active membrane surface and possibly deposit there as a thin gel layer characterized by enzymatic catalytic activity. Actual gelation of enzyme proteins, and hence their dynamic immobilization, depends strictly on enzyme concentration at the membrane–liquid interface.

When the maximum enzyme concentration is lower than the gel concentration value, enzymes are not immobilized. Although they are confined near the membrane surface at fairly high concentration levels, they are still in soluble form (Drioli et al., 1975).

If gel formation occurs, enzymes are effectively immobilized without meaningful changes in their microenvironment. This immobilization is particularly useful owing to the high enzyme and protein concentrations in the gel, the latter strongly enhancing enzyme stability.

Enzyme gel layers can be built up under a number of fluid dynamic conditions. Unstirred and stirred batch reactors have been used as well as systems where the enzymatic solution is kept continuously flowing along semipermeable membranes until gel formation sets in.

When unstirred batch membrane units are used as reacting vessels, a steady-state mass balance on retained species, i.e. enzymes, leads to the evaluation of their concentration as a function of the distance x from the membrane surface. Taking into account both convective and diffusive mass transfer mechanisms, the enzyme mass balance equation at steady state is

$$D_e \frac{d^2 E}{dx^2} + v \frac{dE}{dx} = 0 \tag{2.20}$$

Proper boundary conditions can be derived assuming (a) complete enzyme rejection and (b) no enzyme loss, that is

BC1 $$x = 0 \quad D_e \frac{dE}{dx} + vE = 0 \tag{2.21}$$

BC2 $$\int_0^\infty E(x) A \, dx = N$$

Upon integration of equations (2.20) and (2.21), the enzyme concentration profile appears to be

$$E(x) = [Nv/(AD_e)] \exp[-vx/(D_e)] \tag{2.22}$$

Maximum enzyme concentration occurs at the upstream membrane surface (for $x = 0$) and can be expressed as

$$E_W = \frac{Nv}{AD_e} \tag{2.23}$$

Enzyme depth of penetration (X_e), defined as the ratio of gel volume (in which 99% of the initial amount of enzyme is contained) to membrane cross-sectional area, can be estimated as

$$X_e = 4.6 \frac{D_e}{v} \tag{2.24}$$

where

D_e = enzyme diffusion coefficient $[L^2T^{-1}]$

E = enzyme concentration $[ML^{-3}]$

v = linear velocity $[LT^{-1}]$

A = membrane surface area $[L^2]$

N = total enzyme amount in the reactor $[M]$

With the enzyme gel concentration known, equation (2.22) can be used as a design equation to estimate the amount of enzyme, N, required to build up an enzymatic gel layer of thickness x.

Similar equations under different fluid dynamic conditions can be derived from Michaels' gel formation theory or from similar theories modelling concentration polarization phenomena. Unfortunately, for most substances, an accurate knowledge of gel concentration is not available, and only order-of-magnitude estimates can be gained from the literature. Protein gelation, for instance, is thought to occur at concentration levels ranging from 20% to 40% by weight. Criteria need to be developed for experimental detection of enzyme gel build-up.

Ultrafiltration of protein solutions usually results in a progressive decay of permeate flow rate, which after a period of time attains a steady-state constant value for a given transmembrane pressure. Such a decay can be attributed either to gel formation or to an increase in osmotic pressure of the ultrafiltered solution as a result of macromolecule accumulation at the membrane–solution interface that lowers the pressure driving force. If a gel is formed, further transmembrane pressure increase does not enhance permeating fluxes correspondingly.

Ultrafiltration of an enzyme solution through an UF membrane does not always result in gel layer formation. Unless a gel layer is formed, this immobilization technique cannot be used for flow systems lacking effective enzyme immobilization. In any event, soluble enzyme membrane reactors can be useful for performing kinetic analysis at high enzyme concentrations. Once steady state is attained, the theoretical model permits calculation of reaction rates in both regions. Once the layer is formed it behaves like a secondary membrane, capable of separating compounds of different molecular mass in the mixture as well as catalyzing a chemical reaction.

Enzyme immobilization via dynamic formation of an enzyme gel layer has been applied to both flat and tubular membrane reactors, recirculating either the permeate or the axial stream.

The kinetic behaviour of these immobilized systems has to be analyzed taking into account that: (a) the rate equations for native enzymes and for enzymes in gel form are not necessarily the same owing to microenvironmental and shear stress effects; (b) within the gel layer, substrate mass transfer and reaction occur simultaneously, giving rise to substrate concentration profiles at levels lower than the feed concentration; (c) external mass transfer resistances have to be taken into account occasionally, depending on operating conditions and reactor configuration. Experimental kinetic data at different temperatures and at saturating substrate concentrations can be used to evaluate the relative importance of all these phenomena. Plotting such data as the logarithm of the specific reaction rate vs the reciprocal of the absolute temperature (i.e. an Arrhenius plot) is helpful in assessing which step is rate controlling. A reduction in the activation energy relative to that of the native enzyme indicates that kinetics are not rate limiting.

Comprehensive models of unstirred enzyme gel flat membrane reactors have been proposed (Greco et al., 1979; Drioli et al., 1989).

It is interesting to note that transmembrane pressure plays a different role from that which it plays in the usual UF membrane separators. On the one hand, increasing pressures lead to increasing permeating fluxes, thus enhancing reactor productivity; on the other hand, high effluent flow rates strongly reduce the substrate conversion, with a negative effect on the catalytic activity of the immobilized biocatalyst. This behaviour is reported in Figure 2.14 in terms of specific activity as a function of transmembrane pressure for the case of gelified DNase (Donato et al., 1995).

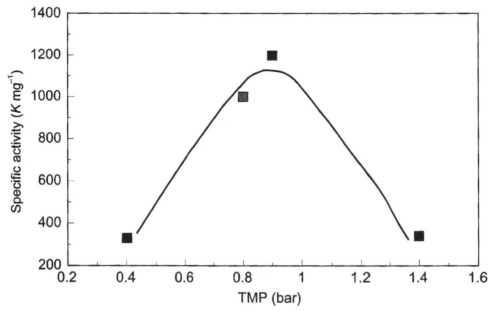

Figure 2.14 Activity of immobilized DNase as a function of transmembrane pressure. (One K unit = a difference in absorbance at 260 nm of 0.001 min^{-1} ml^{-1} at pH 5, 25 °C.)

As for membrane UF units, tubular membranes fitted in cylindrical shells in a 'tube-and-shell' configuration help in improving the performance of enzyme gel reactors. The ratio of filtering area to volume is an order of magnitude higher than that for membranes in a flat slab configuration, and the flow dynamic conditions are more easily controlled. An enzymatic gel layer can be built up on the inner wall of tubular membranes either by filtering proteic solutions in a batch mode or by flushing them along the membrane wall until the gel layer is eventually formed. In both cases, industrial operations require the enzymatic reactor to be operated continuously. Feeding an axial substrate stream to the reactor gives rise to new flow dynamic conditions. High shear stresses may in fact develop on the gel layer surface, leading to partial or total removal of the enzyme. There is evidence of stable systems, at least under laminar flow conditions. Moreover, under a fluid shear field, enzyme molecules can be oriented and thereby denatured. To develop this immobilization technique for large-scale and industrial applications, mechanical stability of the gel layer has to be guaranteed. This can be achieved, first of all, by strictly controlling the axial flow rate. The enzyme gel layer can also be protected by applying layers of water-soluble or insoluble macromolecular compounds or by forming the gel within the porous structure of the membrane where it is less subject to shear stresses.

Flushing the substrate solution along the enzymatic gel causes the substrate to be converted to product even in the axial stream. When the enzyme is product inhibited and the effluent from the reactor is recycled, product accumulates in the feed stream, thus inhibiting gelled enzymes. High axial flow rates may reduce conversion of substrate to product in the axial stream and enzyme inhibition, while product conversion in the permeate remains unaltered at a given transmembrane pressure. Under such conditions, the axial flow rate needs to be optimized since it plays an opposite role (Iorio et al., 1984).

Yeast invertase, acid phosphatase, urease, β-glucosidase, dCMP-aminohydrolase, malic enzyme, DNase and lipase have been immobilized in gel form on both flat and capillary membranes. Cellulosic and polyamide polymers have been used as supporting membrane matrices. In all instances, immobilized enzymes behave in a manner almost identical to their behaviour in homogenous solution, independently of the nature of the polymer. Allosteric or pseudo-allosteric enzymes, proteins whose kinetic behaviour is affected by the presence of particular compounds in the reaction environment (ligands), show no different kinetic behaviour from that when they are subjected to less gentle immobilization procedures.

A number of procedures have been suggested to improve enzyme gel kinetic and mechanical stability. Under ultrafiltration at suitable conditions any proteic solution can give rise to a gel layer. Besides simple enzyme ultrafiltration, two other gel formation procedures have been proposed, namely cogelation and copolymerization/gelation. In the first case, solutions consisting of both enzymes and high molecular mass inert compounds are ultrafiltered through semipermeable membranes. For example, polyalbumins (inert proteins), water-soluble and insoluble compounds have been used as cogelling agents. In the second case, before the UF step, enzymes are chemically linked to high molecular mass inert substances by means of bridge molecules. In both cases, the enzyme microenvironment in the gel layer is characterized by an appropriate protein concentration, close to gelification. Copolymerized/gelled and cogelled enzymatic layers appear to be mechanically stable over a fairly wide range of temperatures and flow rates in the laminar regime (Alfani et al., 1982).

Owing to the covalent linking procedure, copolymerized/gelled enzymes can lose most of their original activity. Furthermore, the deactivation rate of cogelled enzymes appears to be less temperature-dependent than that of copolymerized/gelled preparations. These features suggest that in many cases immobilization by cogelation is preferred, provided that it also gives satisfactory mechanical stability.

The flexibility of enzyme gel layer reactors is fully exploited when multienzymatic reactions are to be performed. It may happen that different enzymes involved in a given transformation cannot be subjected to the same immobilization procedure. Sequential enzyme gel layers can then be built up on the surface of a membrane in the appropriate sequence. Series reactions can be performed in such a set-up, products from one enzymatic layer being fed to the following ones for further transformation. In addition, enzyme gel reactors are inexpensive and easily controlled. Deactivated enzyme may be replaced easily.

Most of the membrane-segregated enzyme systems previously examined suffer some inherent drawbacks that limit their yield and area of application.

2.5   Enzyme-Bound Membrane Reactors

Various other membrane reactors have been studied in which the biocatalysts are immobilized by chemical or physical methods, or on the membrane surface or in the membrane structure itself. These systems are described in the following.

2.5.1   *Membrane Reactors with Chemically Bound Biocatalyst*

Stable attachment can result from ionic binding, cross-linking and covalent linking to a water-insoluble matrix (Messing, 1975).

This section concerns mainly enzymes bound to synthetic polymeric membranes via covalent binding. Since 1954, when protease was covalently bound to diazotized polystyrene, enzyme immobilization via covalent bonds has been an established immobilization

Table 2.2 Enzymes covalently bound to polymers

Enzyme	Polymer
Alcohol dehydrogenase	Nylon
Asparaginase	Nylon
Aldolase	Cellulose, aminoethyl ether
α- and β-Amylase	Cellulose
Asparaginase	Cellulose, cuprofan
Chymotrypsin	Polyacrylamide, cellulose
Glucose-6-phosphate dehydrogenase	Acrylamide
Lactate dehydrogenase	DEAE-cellulose
Lipase	Polyaminostyrene
Penicillin amidase	Cellulose
Catalase	Cuprofan
Allantoinase	Cuprofan
Urease	Cellulose, polysulfone
Papain	Cellulose, polysulfone

technique, usually carried out by means of extremely active bridge molecules, such as CNBr, or bi/multifunctional reagents such as glutaraldehyde. However, the mechanisms involved in enzyme immobilization are not well understood in most cases. When glutaraldehyde is used as a coupling agent, it has been suggested that immobilization results from reaction between the enzyme free amino groups and the glutaraldehyde aldehyde functions with the formation of an intermediate Schiff base. The resultant bonds are generally extremely stable, owing to the high binding energy of the covalent bonds. On the other hand, coupling agent molecules, usually quite small, can penetrate deeply into the active sites of enzyme protein coils where reaction takes place. Once these sites are involved in linking to the matrix, they are no longer available to substrate molecules, resulting in an irreversible loss of activity compared to the initial activity of the native enzymes. When the extent of initial denaturation is acceptable in the economics of the process, enzymes bound to membranes can be used in continuous flow reactors. In Table 2.2 a list of enzymes bound to polymers is reported. Apparently the immobilization procedure affects the immobilized enzyme activity more than the membrane configuration.

Enzymes are covalently immobilized primarily onto the surface of tube membranes exposed to the feed solution, known as the 'active side' of the asymmetric membrane. In general, it is not clear whether reaction between enzymes and polymeric membranes via coupling agents simply results in enzyme attachment to the membrane or leads to an enzyme–carrier network inside the polymer matrix. For the sake of simplicity let us assume that asymmetric membranes are used, that suitable active groups are available on the polymeric surface and that the membrane molecular mass cut-off is such that the active layer is enzyme-impermeable. In this way, even though their activity is often drastically reduced, surface bound enzymes are in close proximity to the substrate solution, thus reducing the mass transfer resistance to that associated with the boundary layer. When enzymes are covalently immobilized in the sponge of the membranes, the mass transfer through the membrane wall must also be taken into account.

Applications of covalently immobilized systems include membrane electrodes for analytical purposes; reactions of substrates whose molecular mass is low compared to membrane molecular mass cut-off; and enzymatic conversion of macromolecules to lower molecular mass species able to permeate the supporting membrane.

When the substrate has a molecular mass lower than the membrane cut-off, enzymes can be covalently bound either to the 'active side' of the membrane or within the sponge-like substructure of the membrane.

If the enzyme kinetic behaviour is not affected by compounds in the solution to be processed, enzymes are preferentially bound onto the 'active side' of the supporting membrane, thus minimizing the overall substrate mass transfer resistance. When such systems are used for analytical purposes, the diffusional resistance in the bulk phase must be taken into account. Obviously, the kinetic parameters for native enzymes are no longer applicable for the immobilized system.

A theoretical analysis of a tubular reactor with impermeable inner walls coated with enzymes was carried out by Kobayashi and Laidler (1974) and experimentally confirmed by Bunting and Laidler (1974).

In addition to applications in industrial processes, enzymes bound to the active side of hollow fibres assembled in a 'tube-and-shell' configuration have been and are under study as extracorporeal or '*in vivo*' devices for use in hepatic failure or to assist leukemic patients. For these therapeutic applications, the irreversible binding of the enzyme to the membrane is extremely important (see Chapter 7) (Callegaro and Rossi, 1981).

When symmetric membranes are used or when enzymes are fed to the spongy part of asymmetric membranes, enzyme immobilization results in either a uniform fixation of enzymes throughout the membrane wall or the formation of a carrier–enzyme insoluble network in the sponge of the membrane. Mass transfer through this solid phase must therefore be taken into account. A theoretical model neglecting radial convective transport and the dense layer in asymmetric membranes is available in the literature (Horvath et al., 1973).

Enzyme immobilization in the sponge of polymeric asymmetric and symmetric membranes has the advantage of a stable immobilized enzyme system along with the improved isolation of enzymatic proteins from immune system actions, high molecular mass inhibitors and proteolytic enzymes. A suitable choice of membranes based on their separation properties allows substrate molecules to permeate through the membrane while, at the same time, separating undesirable compounds from the enzymatic proteins.

The mass transfer mechanisms operative in substrate conversion are essentially those described by Waterland et al. (1974) in their model of the compartmentalized enzyme membrane reactor. Since kinetic parameters cannot be assumed equal to those of native enzymes, a kinetic analysis has to be performed in order to characterize enzyme behaviour after the immobilization procedure.

Sometimes even membrane transport and mechanical properties are affected by the harsh chemical treatments required in the immobilization procedure.

In extracorporeal devices, commercial Cuprophan hollow-fibre membranes are often used. To reduce the side effects of free enzymes in intravenous injection therapies, asparaginase, catalase, uricase, allantoicase and allantoinase have been immobilized onto these membranes, by means of glutaraldehyde or CNBr. *In vivo* fluid dynamic conditions strongly limit the range of conditions at which such reactors can be operated. Reactors are often forced to operate in the diffusion-controlled regime. Under these operating conditions, the overall reaction rate depends on the substrate supply rate; the apparent enzyme activity can thus be increased by increasing the recirculation flow rate. Mazzola and Vecchio (1981) showed that there is a value of the axial flow rate at which the maximum reaction rate is attained; higher flow rates result in a decrease in apparent enzyme activity.

2.6 References

ADU-AMANKWA, B. and CONSTANTINIDES, A., 1984, *Biotechnol. Bioeng.*, **26**, 156–166.

ALFANI, F., ALBANESI, D., CANTARELLA, M. and SCARDI, V., 1982, *Enzyme Microb. Technol.*, **4**, 181–184.

ATKINSON, B. and MAVITUNA, F., 1984, Immobilized enzyme properties, in *Biochemical Engineering and Biotechnology Handbook*, pp. 540–578, The Nature Press, New York.

BAILEY, J.E. and OLLIS, D.F., 1986, *Biochemical Engineering Fundamentals*, 2nd edn, pp. 202–227, McGraw-Hill, New York.

BLATT, W.F., DRAVID, A., MICHAELS, A.S. and NELSEN, L., 1970, in FLINN, J.E. (ed.) *Membrane Science and Technology*, pp. 40–97, Plenum Press, New York.

BRESLAU, B.R. and KELCULLEN, B.M., 1975, Hollow fibre enzymatic reactors. An engineering approach, presented at the *Third International Conference on Enzyme Engineering*, Portland, Oregon.

BUNTING, P.S. and LAIDLER, K.J., 1974, *Biotechnol. Bioeng.*, **16**, 119–134.

CALLEGARO, L. and ROSSI, V., 1981, in *Hollow Fibres and Capillary Membranes in New Separation Processes*, pp. 44–51, C.N.R., Italy.

CHAMBERS, R.P., COHEN, W. and BARICOS, W.H., 1976, *Methods Enzymol*, **44**, 291–317.

CHERYAN, M. and DEESLIE, W.D., 1983, *J. Am. Oil Chem. Soc.* **60**, 1112–1115.

CHERYAN, M. and MEHAIA, M.A., 1983, *Biotechnol. Lett.*, **5**, 519–524.

DALVIE, S.K. and BALTUS, R.E., 1992, Distribution of immobilized enzymes on porous membranes, *Biotechnol. Bioeng.*, **40**, 1173–1180.

DAVIS, J.C., 1974, *Biotechnol. Bioeng.*, **16**, 1113–1122.

DAVIS, E.J., COONEY, D.O. and CHANG, R., 1974, *Chem. Eng. J.*, **7**, 213–225.

DEESLIE, W.D. and CHERYAN, M., 1981, *Biotechnol. Bioeng.*, **23**, 2257–2270.

DONATO, L., GIORNO, L., BASILE, A., DRIOLI, E. and CEDRO, A., 1995, Enzymatic hydrolysis of DNA using membrane bioreactor, *J. Biol. Res.*, **71**(1–2), 13–20.

DRIOLI, E. and CATAPANO, G., 1984, *Chimica Oggi*, **7**(8), 11–16.

DRIOLI, E. and SCARDI, V., 1976, *J. Membr. Sci.*, **1**, 237.

DRIOLI, E., GIANFREDA, L., PALESCANDOLO, R. and SCARDI, V., 1975, *Biotechnol. Bioeng.*, **17**, 1365.

DRIOLI, E., IORIO, G., DE ROSA, M., GAMBACORTA, A. and NICOLAUS, B., 1982, *J. Membr. Sci.*, **11**, 365–370.

DRIOLI, E., IORIO, G. and CATAPANO, G., 1989, Enzyme membrane reactors and membrane fermentor, in PORTER, M.C. (ed.) *Handbook of Industrial Membrane Technology*, pp. 401–481, Noyes Publications, Park Ridge, NJ.

ENGASSER, J.M., 1978, *Biochim. Biophys. Acta*, **526**, 301–310.

ENGASSER, J.M., CAARON, J. and MARC, A., 1980, *Chem. Eng. Sci.*, **35**, 99–105.

GIORNO, L., MOLINARI, R., DRIOLI, E., BIANCHI, D. and CESTI, P., 1995, Performance of a biphasic organic/acqueous hollow fibre reactor using immobilized lipase, *J. Chem. Tech. Biotechnol.*, **64**, 345–352.

GOOSEN, M.F.A., O'SHEA, G.M., GHARAPETIAN, H.M., COHN, S. and SUN, A.M., 1985, Optimization of microencapsulation parameters: semipermeable microcapsules as a bioartificial pancreas, *Biotechnol. Bioeng.*, **27**, 146.

GRECO, G., JR., ALFANI, F., LORIO, G., CANTARELLA, M., FORMISANO, A., GIANFREDA, L., PALESCANDOLO, R. and SCARDI, V., 1979, *Biotechnol. Bioeng.*, **21**, 1421–1438.

HAGERDAL, B., LOPEZ-LEIVA, M. and MATTHIASON, B., 1980, *Desalination*, **35**, 365–373.

HONG, J., TSAO, G.T. and WANKAT, P.C., 1981, *Biotechnol. Bioeng.*, **23**, 1501–1516.

HORVATH, C., SHENDALMAN, L.H. and LIGHT, R.T., 1973, *Chem. Eng. Sci.*, **28**, 375–388.

HORVATH, C. and ENGASSER, J.M., 1974, *Biotechnol. Bioeng.*, **16**, 909.

HOWELL, J.A., KNAPP, J.S. and VELICANGIL, O., 1978, in BROUN, G.B., MANECKE, G. and WINGARD, L.B. JR. (eds) *Enzyme Engineering*, Vol. 4, pp. 267–271, Plenum Press, New York.

IORIO, G., CATAPANO, G., DRIOLI, E., ROSSI, M. and RELLA, R., 1984, *Ann. N.Y. Acad. Sci.*, **434**, 123–126.

KLEIN, E., 1990, *Affinity Membranes*, Wiley, New York.

KOBAYASHI, T. and LAIDLER, K.J., 1974, *Biotechnol. Bioeng.*, **16**, 77–97.

LEUCHTENBERGER, W., KARRENBAUER, M. and PLOCKER, U., 1984, Scale up of an enzyme membrane reactor process for the manufacture of L-enantiomeric compounds, presented at the *Europe–Japan Congress on Membranes and Membrane Technology*, Stresa, Italy.

LEVENSPIEL, O., 1972, *Chemical Reaction Engineering*, Wiley, New York.

LEWIS, W. and MIDDLEMAN, S., 1974, *AIChE J.*, **20**, 1012–1014.

LI, A. and SUTTON, P.M., 1982, *Dorr-Oliver's Membrane Anaerobic Reactor System*, Internal Report, Dorr-Oliver.

MATHIASON, E. and SIVIK, B., 1980, *Desalination*, **35**, 59–103.

MAZZOLA, G. and VECCHIO, G., 1981, in *Hollow Fibres and Capillary Membranes in New Separation Processes*, pp. 52–58, C.N.R., Italy.

MEHAIA, M.A. and CHERYAN, M., 1984, *Enzyme Microb. Technol.*, **6**, 117–120.

MESSING, R.A. (ed.) 1975, *Immobilized Enzymes for Industrial Reactors*, pp. 79–99, Academic Press, New York.

MICHAELS, A.S., 1968, *Chem. Eng. Progr.*, **64**, 31–43.

MOSBACH, K., 1980, Immobilized enzymes, *TIBS*, January 1–3.

OHLSON, I., TRAGARDH, G. and HAHN-HAGERDAL, B., 1984, *Biotechnol. Bioeng.*, **26**, 647–653.

PORTER, M.C. and MICHAELS, A.S., 1970, *Chem. Tech.*, **2**, 56–61.

REACH, G., JAFFRIN, M.Y. and ASSAN, R., 1982, *Life Support Systems*, **1** (supplement), 73–75.

REACH, G., JAFFRIN, M.Y. and DEAJEUX, U.F., 1983, A U-shaped flat membrane bio-artificial pancreas. Application of the ultrafiltration-based model to its design, *Proc. Eur. Soc. Artificial Internal Organs*, pp. 49–52.

RONY, P.R., 1971, *Biotechnol. Bioeng.*, **13**, 431–447.

SHIMIZU, Y., SHIMODERA, K. and WATANABE, A., 1993, Cross-flow filtration of bacterial cells, *J. Ferment. Bioeng.*, **76**(6), 493–500.

STRACHAN, L.F. and LIVINGSTON, A.G., 1997, The effect of membrane module configuration on extraction efficiency in an extractive membrane bioreactor, *J. Membr. Sci.*, **128**, 231–242.

TAKABAKATE, H., KOIDE, N. and TSUJI, T., 1991, Encapsulated multicellular spheroids of rat hepatocytes produce albumin and urea in a spouted bed circulating culture system, *Artif. Organs*, **15**(16), 474–480.

TELO, J.P., CANDEIAS, L.P., EMPIS, J.M.A., CABRAL, J.M.S. and KENNEDY, J.F., 1990, Immobilized enzymes — active, partially active or inactive? ESR studies of papain immobilization, *Chemistry Today*, October, 15–18.

TIMMER, J.M.K. and KROMKAMP, J., 1994, Efficiency of lactic acid production by *Lactobacillus helvetica* in a membrane cell recycle reactor, *FEMS Microbiol. Rev.*, **14**, 29–38.

TREVAN, MICHAEL D., 1981, *Immobilized Enzymes*, Wiley, New York.

VICK ROY, T.B., MANDEL, D.K., DEA, D.K., BLANCH, H.W. and WILKE, C.R., 1983, *Biotechnol. Lett.*, **5**, 665–670.

WANDREY, C., 1983, Enzyme membrane reactor systems, in *Proc. Int. Conf. Commercial Applications and Implications of Biotechnology*, pp. 577–588, Online Publications Ltd., Northwood, UK.

WANDREY, C., WICHMANN, R., LEUCHTENBERGER, W., KULA, M. and BUCKMANN, A., 1982, U.S. Patent 4,304,858.

WATERLAND, L.R., MICHAELS, A.S. and ROBERTSON, C.R., 1974, *AIChE J.*, **20**, 50–59.

WICHMANN, R., WANDREY, C., BUCKMANN, A.F. and KULA, M.R., 1981, *Biotechnol. Bioeng.*, **23**, 2789–2802.

WISNIEWSKI, J., WINNICKI, T. and MAJEWSKA, K., 1983, *Biotechnol. Bioeng.*, **25**, 1441–1452.

Appendix 2.1   Immobilized Enzyme Kinetics

The observed catalytic properties of immobilized enzymes depend upon the transport of reagents and the catalytic activity of the enzyme.

The Michaelis–Menten (M-M) constant (K_m) is often larger for an immobilized enzymatic system than for the corresponding homogeneous system. We will consider how the effect of diffusional resistance tends to increase the apparent value of K_m. A stagnant film, referred to as a Nernst diffusion layer, forms around the particle through which the substrate molecules must diffuse to reach the surface. Thus, the surface concentration is less than the bulk concentration; the reaction proceeds less rapidly than the bulk substrate concentration would suggest.

The Michaelis–Menten equation

$$(-r_s) = \frac{V_{max}[S]}{K_m + [S]}$$

(where r_s = initial reaction rate) shows that an increase of K_m also corresponds to a decreased reaction rate; thus, the diffusional effect is not surprising. Diffusional effects can be substantially reduced by increasing the stirring rate and decreasing the size of the suspended particles.

In the 1930s, Thiele, Damköhler and Zeldovitch studied the influence of *diffusion within porous catalysts* upon reaction kinetics. Their techniques are used for the analysis of enzymes entrapped in a matrix or otherwise bound in a porous medium. Consider the example of a planar membrane containing enzyme distributed uniformly throughout. The result of combining the steady-state diffusion equation (with a source term) with the applicable kinetics rate expression, say the Michaelis–Menten equation, is

$$D_{eff} \frac{d^2[S]}{dx^2} - \frac{V_{max}[S]}{K_m + [S]} = 0$$

where

D_{eff} = the effective diffusivity $[L^2T^{-1}]$

$[S]$ = the substrate concentration $[ML^{-3}]$

V_{max} = the maximum velocity of the reaction $[MT^{-1}L^{-3}]$

K_m = the Michaelis constant $[ML^{-3}]$

The values V_{max} and K_m are those that characterize the kinetic expression for the microenvironment within the porous structure.

Suppose that the membrane lies between the planes $x = -L$ and $x = +L$ and that the substrate concentration at these planes, which represent the interfaces between the membrane and the bulk fluid, equals [E]. The equation is cast into dimensionless form by defining (Greene, E.R., private communication; Bailey and Ollis, 1986)

$$\psi = \frac{[E]}{K_m} \quad \text{and} \quad z = \frac{x + L}{L}$$

The resulting system is

$$\frac{d^2\psi}{dz^2} - \phi^2 \frac{\psi}{1 + \psi} = 0$$

$$\psi = \psi_0 \quad \text{at} \quad z = 0$$

$$\frac{d\psi}{dz} = 0 \quad \text{at} \quad z = 1$$

where ϕ, the Thiele modulus, is given by

$$\phi = L \left(\frac{V_{max}}{D_{eff} K_m} \right)^{1/2}$$

The Thiele modulus has the physical meaning of a *reaction rate / diffusion rate*. Horvath and Engasser solved this system numerically for various values of the parameters ψ_0 and ϕ to obtain internal concentration profiles. For each profile they computed the overall reaction rate V within the entire membrane. For $\phi \leq 1$ they showed that the reaction is essentially controlled by kinetics. The normalized overall reaction rate V/V_{max} decreases as ϕ increases, for given ψ_0. This indicates increasing diffusion limitations.

For sufficiently large values of ψ_0, that is for $[S]_0 \gg K_m$, V approaches V_{max}, regardless of the value of ϕ. On the other hand, for small values of ψ_0, that is for $[S]_0 \ll K_m$, $[S]$ must be much less than K_m throughout the membrane. This corresponds to the limiting first-order approximation to the Michaelis–Menten equation, namely

$$(-r_s) = \frac{V_{max}[S]}{K_m}$$

Solving this limiting equation analytically leads to $\psi = \psi_0 \cosh(\phi - \phi z)/\cosh \phi$. Using the usual definition of the effectiveness factor η, that is, the ratio of the actual reaction rate to the maximum possible rate in the absence of diffusion, for small ψ_0, $\eta = (\tanh \phi)/\phi$. As a result

$$(-r_s) = V_{max}[S]_0 \frac{\tanh \phi}{K_m \phi}$$

The use of Lineweaver–Burk plots with immobilized enzymes to determine apparent values for V_{max} and K_m is very common. For a kinetically controlled reaction, a straight line is obtained. A linear extrapolation to the axes allows the apparent values of these parameters to be determined accurately. In the absence of effects directly related to the binding process, such as steric hindrance or severe enzyme modification, or of micro-environmental effects such as the charging of the carrier, the apparent values of K_m and V_{max} will equal those for the soluble enzyme. In the presence of diffusional limitations, the Michaelis constant's apparent value obtained by extrapolation of the Lineweaver–Burk plot is usually larger than that for the soluble form. Under these conditions the M-M equation can no longer account for the overall kinetics.

Mass Transfer from the Bulk to a Surface Coated with a Layer of Enzyme

Let us assume that M-M kinetics apply, although other kinetics schemes can be analyzed in an analogous fashion. At steady state, the rate of substrate transport to the surface equals its rate of consumption by the reaction. Thus

$$k_m a_m([S]_b - [S]_0) = \frac{V_{max}[S]_0}{(K_m + [S]_0)}$$

where a_m is the area of mass transfer and k_m is the mass transfer coefficient. The subscripts b and 0 refer to the bulk and surface concentrations. For flow through a packed bed of spherical particles, K_m can be obtained from the Chilton–Colburn correlation:

$$J_d = \frac{k_m \rho}{G} \left(\frac{\mu}{\rho D} \right)^{2/3}$$

where J_d is a function of the Reynolds number

$$N_{Re} = \frac{d_p G}{\mu}$$

In these expressions, μ and ρ are the kinematic viscosity and density of the fluid, respectively, D is the diffusivity of the substrate, d_p is the particle diameter, and G is the superficial mass velocity of feed to the reactor.

McCume and Wilhelm have shown that

$$J_d = 1.625 \, N_{Re}^{-0.507} \qquad \text{for} \qquad 8 < N_{Re} < 20$$

and

$$J_d = 0.687 \, N_{Re}^{-0.507} \qquad \text{for} \qquad 120 < N_{Re} < 1300$$

By putting the rate equation into dimensionless form by defining ψ:

$$\psi = \frac{[S]}{K_m}$$

gives it

$$\psi_b - \psi_0 = \frac{N_{Da} \psi_0}{1 + \psi_0}$$

where the Damköhler number N_{Da} is

$$N_{Da} \frac{V_{max}}{K_m k_m a_m}$$

and K_m is the M-M constant, k_m is the diffusion coefficient. When N_{Da} is small, ψ_0 practically equals ψ_b and the reaction proceeds at its maximum possible value (V_{kin})

$$(-r_s) = \frac{V_{max}[S]_b}{K_m + [S]_b}$$

and is kinetically controlled. As N_{Da} increases, ψ decreases from ψ_b towards its lower limiting value of zero. In this limit (V_{diff})

$$(-r_s) = k_m a_m [S]_b$$

and the reaction is diffusion controlled.

The two limiting expression for $(-r_s)$ are represented by V_{kin} and V_{diff}. The Damköhler number can be interpreted as the ratio of the slope of V_{kin} to that of V_{diff}, evaluated by plotting $(-r_s)$ vs $[S]_b$. If the slope of $V_{diff} (= k_m a_m)$ is less than that for $V_{kin} (= V_{max}/K_m)$ at $[S]_b = 0$, then $N_{Da} [= (V_{max}/K_m)/(k_m a_m)] > 1$ and the reaction is diffusion controlled; in the opposite case, when $N_{Da} < 1$, the reaction is kinetically controlled.

The subject is treated in greater detail in chemical reaction engineering texbooks (Levenspiel, 1972; Bailey and Ollis, 1986).

Applications of Biocatalytic Membrane Reactors

3

Catalytic Membrane Reactors in Integrated Processes for Production of Bioactive Compounds

3.1 Introduction

The potential for using integrated membrane processes combining catalytic membrane reactors with other separation techniques, such as crossflow microfiltration, ultrafiltration, reverse osmosis, electrodialysis, liquid membranes, and so on, for the production of bioactive compounds is huge. The synergetic effects obtainable by designing the overall biotechnological process in combination with various membrane techniques are particularly significant.

The use of membrane reactors combined with other membrane separation processes is necessary when the product needs further processing, such as concentration, fractionation, purification, and so on. These treatments are particularly important for products obtained by fermentation processes — organic acids, antibiotics, etc. — and in the processing of food and beverages, such as fruit juices, milk or wine.

As examples, the production of carboxylic acids by continuous membrane fermentation and their simultaneous downstream processing by integrated membrane operations is of particular interest. The depectinization of fruit juices in enzyme membrane reactors and subsequent treatment by integrated processes to produce concentrated pulps and juices enriched in aroma compounds, involves attractive new production systems, as also does the hydrolysis of whey and milk proteins, milk sterilization, fractionation, or dehydration (Coca et al., 1992).

The sterilization, separation, purification and concentration of thermosensitive bioactive compounds have been carried out by combining ultrafiltration, ion exchange and reverse osmosis, with significant reduction in costs and increase in productivity (Drioli, 1986). Enzyme membrane reactors have been used in high-temperature lactose hydrolysis, using a thermophilic, thermostable enzyme ultrafiltration membrane (Drioli et al., 1982). Separation of antibiotics from fermentation broths has been carried out by ultrafiltration and membrane-based solvent extraction (Hefner and Giorno, 1992).

The term downstream processing in biotechnology refers to the chain of unit operations that are combined into a system for the recovery, purification and concentration of the products at the lowest possible cost and with recovery at the highest possible quality factor. The recovery step generally represents a large part of the overall capital investment

Table 3.1   Uses of bioactive compounds

Product category	Application
Carboxylic acids	Food industry
Amino acids	Food industry
Vitamins	Pharmaceutical industry
Antibiotics	Pharmaceutical industry
Anti-inflammatory	Pharmaceutical industry
Enzymes	Biocatalytic processes
	Pharmaceutical industry
Hormones	Pharmaceutical industry
Nucleotides	Pharmaceutical industry
Peptides	Pharmaceutical industry
Aromatics	Food industry

in a fermentation plant and its cost efficiency is a key factor in the production of bio-technological compounds and in the development of new bioprocesses.

The recovery of bioactive materials from the fermentation broth is generally complicated by the fact that the bioproducts are in very low concentration, often in unstable non-Newtonian fluids. The optimization of downstream processing is a key area for further development of biotechnology. Membrane techniques can be considered as broad core technologies in this industrial sector. They can contribute to improving the cost-effectiveness of the systems and preserve the product quality, since they operate under mild conditions.

The most important bioactive compounds in the pharmaceutical, biotechnology and food industries are summarized in Table 3.1.

In this chapter, the use of catalytic membrane reactors integrated with membrane operations for bioprocessing of fermentation products (carboxylic acids, antibiotics), wine, fruit juices, milk and whey will be discussed. Further cases of pharmaceutical and bio-technological applications of catalytic membrane techniques will be discussed in the following chapters.

3.2   Bioprocessing of Fermentation Products

3.2.1   *Traditional Technologies for Production and Separation of Carboxylic Acids*

Carboxylic acids are an important group of additives having extensive uses in the food industry (Table 3.2). Such compounds include citric, malic, lactic, tartaric and gluconic acids. All these acids are aliphatic hydroxy acids which differ in their chemical behaviour and organolectic properties.

Citric and malic acids are the principal food acids. They are used to adjust the acid flavour in soft drinks, fruit juices, canned fruit, cider and wine. Citric, lactic and malic acid inhibit the development of metal-catalyzed off-flavours and colour deterioration. Lactic acid might be considered the appropriate material for adjusting acidity in wines. Calcium lactate is a so-called protein plasticizer and is employed in the preparation of

Table 3.2 Uses of carboxylic acids

Substrate	Catalyst	Carboxylic acids produced	Application	Reference
Beet and molasses, glucose, sucrose	*Aspergillus niger*	Citric acid	Food acid (improve flavour of foods)	Arnold (1975)
Molasses, glucose	*Aspergillus flavus Aspergillus oryzae Pseudomonas*	Malic acid	Food acid acidulant	Morisi et al. (1976)
Glucose, beet molasses, sucrose, whey, hydrolyzed starch	*Aspergillus delbrukii bulgaricus*	Lactic acid	Food preservative	Samuel (1980)
Glucose syrup	Gluconobacter suboxydans *Aspergillum niger*	Gluconic acid	Foodstuff	Nilson (1986)

dried milk powders for baby food and also in foamed protein products. Lactate polymers can be used as benign packaging thanks to their biodegradability. The demand for environmentally benign packaging products has reduced the production of litter and pollution. Because of their high strength, thermoplasticity, fabricability, biodegradability, bioenvironmental compatibility and availability from renewable resources, lactate polymers are receiving increasing attention.

All of these applications show the importance of carboxylic acids and the possibility of increasing the world market of these compounds provided more efficient and cost-effective processes can be developed.

The approximate annual production tonnages of the carboxylic acids are as follows (Milsom, 1986): citric, 350 000 tonnes/year; malic, 25 000 tonnes/year; tartaric, 40 000 tonnes/year; gluconic, 50 000 tonnes/year; and lactic, 20 000 tonnes/year.

Carboxylic acids are traditionally produced in batch fermentors. Depending on the kind of medium and microorganism, tower fermentors, stirred fermentors or rotor fermentors (Margaritis and Wilke, 1978) have been used. The carboxylic acids obtained were usually separated at the end of the fermentation process. The separation occurred through a sequence of discontinuous operations, such as filtration (to remove cells and solid materials); acidification (addition of sulfuric acid to precipitate calcium sulfate); preliminary evaporation (to concentrate the diluted acid); removing of metals; further evaporation; crystallization; etc.

Fermentation processes for obtaining citric acid (2-hydroxy-1,2,3-propane tricarboxylic acid, $HOOC-CH_2-C(OH)COOH-CH_2-COOH$) employed the mould *Aspergillus niger* growing on the surface of a medium consisting of a solution of sucrose and inorganic salts. Further developments in this processes led to the use of beet molasses as the main constituent of the medium. The surface and submerged processes are still used (Milsom, 1986). The surface process requires more manpower but less energy than the submerged method. Advantages of the submerged method include the possible use of a wider range of substrates and better control of the fermentation. Substrates used include glucose, sucrose, beet molasses and cane molasses. Other sources are whey permeate (Hossain et al., 1983) and apple pomace (Hang and Woodams, 1984). Sucrose, glucose, fructose and lactose are the best substrates for citric acid production.

In recent years it has been found that certain species of yeast cultivated on carbohydrate media are able to accumulate citric acid. The advantages of the yeast-based process are said to be high productivity and relative insensitivity to variations in molasses.

In the classical recovery process for citric acid, clarified fermented liquor is treated with lime to precipitate the insoluble tricalcium salt. The precipitate is carefully washed to remove soluble impurities and treated with sulfuric acid to precipitate calcium sulfate and regenerate the citric acid. The solution so obtained is concentrated, treated with ion exchangers and charcoal and finally crystallized. Evaporative crystallization yields the anhydrous material, whereas cooling crystallization gives monohydrate. Further crystallization from water produces a high-grade material suitable for food use.

Solvent extraction methods are also suitable for the recovery of citric acid, in particular when the fermentation medium is composed of sucrose or glucose and inorganic salts. Media based on molasses contain solvent-extractable impurities. Surface-active components of molasses make the separation of the solvent and aqueous phases rather difficult.

Lactic acid (2-hydroxypropionic acid) (CH_3-CHOH-COOH) is produced from mono- or disaccharides via the Embden–Meyerhof glycolysis (Lehninger, 1987). The properties of lactic dehydrogenase have been studied in the bacterium *Lactobacillus plantarum*, which produces a recemic mixture of the L(+) and D(−) stereoisomers. This bacterium elaborates two stereospecific lactate dehydrogenases, one producing the L(+) isomer and the other the D(−) form. Under the influence of these two enzymes, racemization of lactic acid can take place. Organisms that produce only one stereoisomer of lactic acid must contain only one active lactate deydrogenase and no active racemase.

In the production of lactic acid by anaerobic fermentation, strains of the so-called homolactic organisms such as *Lactobacillus delbruckii* or *Lactobacillus bulgaricus* are usually used. These organisms form only lactic acid from the carbohydrate supplied. The choice of medium in the lactic acid fermentation has a great effect on the process as a whole. It is well known that the recovery of lactic acid from the fermentation broth is more difficult than the fermentation. The cheapest raw material (beet molasses) causes the most problems in recovery, whilst the most expensive (sucrose) gives the least. In the United States, glucose or enzyme-hydrolyzed starch are the most commonly used substrates. In Europe, both molasses or sucrose have been employed. Other substrates can be whey and cellulose.

Lactobacillus needs a medium pH between 5.5 and 6.0 in order to ensure a satisfactory rate of fermentation. At pH lower than 4.5, no lactic acid is produced. The fermentation lasts between 3 and 5 days, depending on the organism, the medium and the conditions. The highest yield of lactic acid ranges between 80% and 90% of the carbohydrate supplied.

Lactic acid can be obtained in a pure crystalline form melting at about 54 °C. However, special methods are needed to obtain the crystalline material.

Because lactic acid contains both a carboxyl and hydroxyl group, it can condense with itself to form esters such as lactolyl-lactic acid and dilactide (Holten, 1971). When a solution of lactic acid is evaporated, part of the lactic acid is converted to these compounds and to polymers containing more than two lactic acid units, so that a somewhat viscous noncrystallizing syrup is obtained. At the end of the fermentation carried out with calcium carbonate as a pH controlling agent, the bacteria and any other solid material are removed by filtration, giving a solution of calcium lactate. Sulfuric acid is added and the calcium sulfate so formed is removed by filtration. If the fermentation medium has been based on sucrose and this has been substantially consumed, recovery is relatively easy. After a preliminary evaporation, colouring matters may be removed. If any iron is present, giving an off-colour, it must be precipitated. Finally, the solution may be concentrated again by evaporation to give a food-grade material containing about 80% lactic acid. A higher purity material may be obtained by extracting the partly evaporated lactic acid solution with a solvent, followed by re-extraction into water.

If a crude carbohydrate, e.g. molasses, is used in the fermentation medium, it becomes essential to use a solvent extraction and/or an esterification procedure to obtain a satisfactory food-grade product; that is, the lactic acid is reacted with methanol or ethanol and the obtained ester is purified by distillation and then hydrolyzed to give the free pure lactic acid.

Purification of lactic acid may also be carried out using ion exchangers (Buchta, 1974). Other methods for the separation of lactic acid from filtered, decalcified fermentation broth include reverse osmosis (Matsuura, 1973) and electrodialysis.

The downstream separation represents the most expensive part in the production of carboxylic acids. The recovery of carboxylic acid from fermentation broth is complicated by the very low concentration in the fermentation broth solution. For these reasons, today most work in this field is oriented to the development of integrated systems for continuous production and downstream processing of substances produced.

In the following sections the evolution of methods for the production and separation of carboxylic acids is discussed.

3.2.2 *Innovative Integrated Membrane Processes for Production and Recovery of Carboxylic Acids*

Although batch fermentations appear to dominate the industry today, there is much interest in developing more efficient methods of producing compounds of interest in a continuous way.

In citric acid production, a continuous process in which fresh medium is continuously added to the fermentor, and a mixture of cells and fermented liquor is withdrawn at the same rate, would avoid the time-consuming start-up phase of batch fermentation. In practice, however, such continuous processes are best suited to a directly growth-related product (such as with lactic acid).

Another possibility for obtaining more efficient citric acid production is the immobilized enzyme reactor. In an idealized system, a series of immobilized enzyme reactors would carry out the conversion of sucrose to citric acid. But some of the enzymes are very unstable and/or need cofactors for their operation. Such cofactors could be lost to the substrate or need to be continually regenerated (new membrane techniques for this purpose will be also discussed in Chapter 5).

The use of immobilized cells that contain all the necessary enzymes, cofactors and cofactor regeneration systems would be more appropriate. Cells have been immobilized by adsorption on solid supports, entrapment in gel matrices, and covalent attachment to inert supports.

Various supports have been used. Borglum and Marshall (1984) studied *Aspergillus niger* cells entrapped in beads either of agar or of k-carrageenan gel. Citric acid was produced from glucose using these gels but at a reduced rate compared with the performance of *Aspergillus niger* pellets. Vaija et al. (1982) immobilized pellets of *Aspergillus niger* in calcium alginate gel beads and studied the production of citric acid in a continuous system. Eikmeier and Rehm (1984) immobilized spores of *Aspergillus niger* in beads of calcium alginate and allowed these to grow by immersing the beads in a growth medium. In order to prevent outgrowth of mycelium from the beads, the growth medium was nitrogen limited. A 50% yield of citric acid on sugar consumed was obtained.

Tuli et al. (1985) immobilized cells of *Lactobacillus casei* in agar and polyacrylamide. A temperature range of 40–50 °C and a pH range of 4.5–6.0 were established as optimal. 90% conversion of lactose to lactic acid was observed.

The use of membranes and membrane techniques for the production and separation of carboxylic acids will be discussed in the following section.

Many studies have been directed at the realization of continuous fermentations using hybrid systems of fermentation and molecular filtration. With such systems, the product is first separated from cell debris and macromolecules, and it needs to be separated further from substrate and nutrients with low molecular mass. Membrane technologies (electrodialysis, pertraction, nanofiltration, etc.) have been investigated and are currently under development for continuous downstream processing of carboxylic acid. These new technologies will be discussed in the following sections. Studies on various carboxylic acids are reported, but lactic acid is probably the best-known case, followed by citric acid. For this reason, the technologies presented here will refer basically to these two cases.

Production of carboxylic acids in CMRs such as continuous membrane fermentors

Hybrid set-ups of traditional fermentor combined with membrane filtration processes have been widely studied during the 1990s. In the literature they are variously referred to as 'cell recycle membrane (bio)reactors' (CRMR), 'cell recycle membrane fermentors' (CRMF), continuous membrane fermentors (CMF), and other names. All of them have the same basic concept of catalytic membrane reactors, which combine chemical conversion (in this specific case a fermentation) with the separation of the product by membrane operations.

The developments of these new fermentation technologies showed increased productivity compared to batch fermentation (Cheryan and Mehaia, 1985; Mehaia and Cheryan, 1987; Taniguchi et al., 1987; Ohara et al., 1992, 1993; Shimizu et al., 1993).

A continuous membrane fermentor is described schematically in Figure 3.1. The system consists of a fermentor connected to an UF-membrane module. The system as a whole can be seen as a catalytic membrane reactor in which the biocatalyst is compartmentalized in the bulk reaction by the membrane module. Cells, macromolecules and solids are retained by the membrane and recirculated to the fermentor. Product and other small molecules permeate through the membrane and are removed from the fermentor. The volume of the fermentor is kept constant by adding fresh feed at the same rate as the permeate flow rate. The product is diluted in the fermentor solution; concentration is reduced and product inhibition is avoided or at least limited. Since the substrate (glucose, sucrose) and low molecular mass nutrients permeate through the membrane together with the product, it is convenient to start the ultrafiltration process when the concentration of nutrients is decreased and that of product is increased. It must be taken into account that the concentration of substrate should not be lower than the limiting value for the survival of cells.

Since in a continuous fermentation mode the microorganism is continuously fed it grows and its density is increased considerably. This fact allows the reactor to operate at high catalyst density. On the other hand, high density and viscous solutions are difficult to treat by membrane processes. Furthermore, the fouling created by the macromolecules decreases membrane separation efficiency. An optimal cell density must be maintained in the fermentor by using an appropriate bleed rate. This means that part of the fermentation broth has to be removed and replaced with fresh medium at regular intervals.

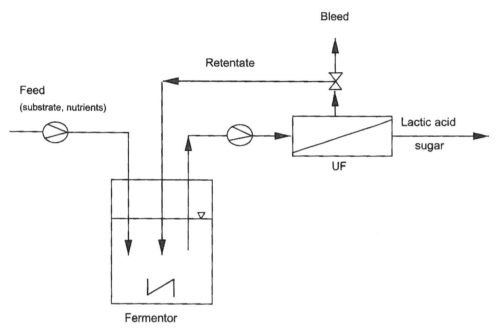

Figure 3.1 Schematic of a continuous membrane fermentor set-up.

Table 3.3 Comparison between characteristics of continuous and batch fermentors

Continuous membrane fermentor	Batch fermentor
Continuous operation	Start-up and shut-down procedures
Reuse of biocatalyst	High volume requirement
Low product inhibition	Product inhibition
High cell density	Low cell density
High productivity	Low productivity
Product in clear solution	High labour costs
Reduced waste	High product concentration at the end of the process
Low labour costs	
Low product concentration in the permeate	
No product separation from substrate and nutrients	

The process variables in a continuous membrane fermentor are:

Flow rate of fermentation broth, Q ($1\,h^{-1}$)

Dilution rate, $D = Q_p/V$ (h^{-1}); where Q_p = permeate flow rate, V = fermentation volume

Bleed rate, B (h^{-1})

Starting time of ultrafiltration after onset of fermentation.

These parameters have to be optimized case by case. Iwasaki et al. (1993) showed that the lactic acid production rate is related to the dilution rate; in particular, the production rate increases with increasing the dilution rate. They also showed that the productivity of continuous fermentation was higher than that of batch fermentation.

A comparison between characteristics of continuous membrane fermentors and batch fermentors is reported in Table 3.3 and an overview of carboxylic acid production using different microorganisms in continuous membrane fermentors (CMFs) is given in Table 3.4. Productivity is in general higher than that of batch fermentor processes. However, final lactic acid concentration in the permeate is lower than in the case of batch fermentation.

The economic advantage of a high productivity process decreases rapidly when product concentration decreases. Lactic acid production by continuous membrane fermentor is interesting on an economic basis when production is on a relatively small scale and the lactic acid concentration is as high as possible. The main reason for this is the high cost of water removal.

Continuous membrane fermentors are realized by combining a traditional fermentor with crossflow microfiltration, ultrafiltration, or electrodialysis membrane units. Crossflow microfiltration performance is affected by operating parameters, fluid characteristics and membrane properties.

In ultrafiltration and nanofiltration theories (Fane et al., 1981), the gel polarization model (Nakao et al., 1979) and the osmotic pressure model (Wijimans et al., 1984), have been proposed to relate the operating parameters (transmembrane pressure and feed velocity) and fluid characteristics (concentration of solute, molecular mass of solute, and temperature) to flux. The applicability of the flux prediction equation to biotransformations has been investigated by Shimizu et al. (1993) using the crossflow filtration of baker's yeast as a model case. They studied the influence of operating parameters and fluid characteristics on the steady-state flux during crossflow microfiltration and found an exponential relationship between the steady-state flux and various factors such as feed velocity, particle concentration, particle size and viscosity.

Moueddeb et al. (1996) proposed a tubular membrane bioreactor containing two co-axial porous alumina tubes for the transformation of lactose into lactic acid by *Lactobacillus rhamnosus*. The microorganism is fixed in the macroporous support and confined in the space defined by the two separating walls. The substrate solution (200 ml) is fed into the inner compartment and permeates in the radial direction across the two membranes. An initial solution of 28 g l^{-1} of lactose was completely converted into lactic acid after 90 h of continuous operation. The reactor, in the indicated conditions, needs to work continuously for a relatively long time in order to convert the substrate completely. Lengthy contact time and high microorganism concentration are necessary to have good substrate conversion. They also developed a model in cylindrical coordinates that takes into account the mass transfer phenomena coupled with biological reaction in the membrane annulus.

Taniguchi et al. (1987) showed that in a continuous crossflow membrane fermentor the productivity of lactic acid was 19-fold compared with the corresponding conventional batch cultivation, thanks to reduced product inhibition and high cell density.

Within the framework of a European research project (Copernicus ERB-CIPA CT92-3018), a continuous cell recycle membrane fermentor for production of lactic acid has been studied. One-litre or 2.5 litre batch fermentors were used. MRS medium containing 20 g l^{-1} of glucose and other nutrients was inoculated by using 2.5% by volume of 24 h and 48 h culture of *L. bulgaricus*. Conversion of glucose was completed in about 20 h with an average yield $[(g_{lactic\ acid}/g_{glucose}) \times 100]$ of about 90%.

The selection of UF membrane for the set-up of a continuous cell recycle membrane fermentor was made on the basis of flux and retention properties determined with fermentation broth, at operating conditions of flow rate 160 l h^{-1} and transmembrane pressure (TMP) 5×10^4 Pa. Polyamide membranes with 50 kDa cut-off have shown the best performance (Figure 3.2a).

Table 3.4 Technical aspects of CMFs producing carboxylic acids

Catalyst	Medium	Product	Membrane reactor	Membrane operation	Immobilization	Productivity (g l⁻¹ h⁻¹)	Product concentration in the permeate (g l⁻¹)	Product output form	Reference
Lactobacillus bulgaricus	Whey permeate Yeast extract	Lactic acid	CRMR	UF	Compartmentalized in the fermentor	84	117	In solution with other components	Mehaia and Cheryan (1987)
L. delbruckii	Lactose	Lactic acid		UF	Compartmentalized	66	69	In solution with other components	Ohleyer et al. (1985)
L. helveticus	Whey permeate Whey concentration	Lactic acid	(no biomass) CRMR	UF	Compartmentalized	37.5	50	In solution with other components	Timmer and Kromkamp (1994)
L. rhamnosos	Lactose and nutrients	Lactic acid	Tube-in-tube membrane reactor	MF (ceramic)	Fixed in the macroporous and confined in M.R.	~0.3	~30	In solution with other components	Moueddeb et al. (1996)
Saccharomyces cremoris L. casei	Lactose and nutrients	Lactic acid	CRMR	MF (ceramic tube and multitube)	Compartmentalized in the fermentor	0.9	15 + 20	Lactate in solution with other components	Taniguchi et al. (1987)
L. casei	Glucose Yeast extract	Lactic acid	CRMR	ED	Compartmentalized in the fermentor			In solution with other components	van Nispen and Jonker (1991)

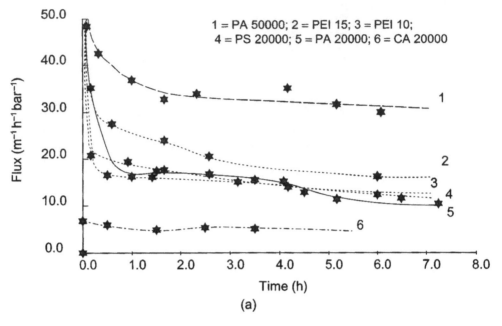

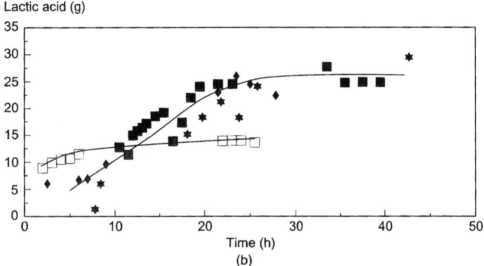

Figure 3.2 (a) Performance of ultrafiltration membranes using fermentation broth. (b) Total amount of lactic acid produced by continuous (■, ◆, ✱) and batch fermentations (☐).

From batch fermentation experiments, it was observed that the growth of bacteria, and the production of lactic acid, did not start before 5–7 h. These results suggested that it would not be useful to run the ultrafiltration during this period, because of the low cell density and lactic acid concentration. On the other hand, it is necessary to feed the fermentor before the concentration of substrate decreases below the survival limit of bacteria. The optimization of this parameter has been carried out by starting the ultra-filtration process at different times after the onset of the fermentation. At the steady state, the operating dilution rate was about 0.1 h^{-1}, the feed flow rate about 160 l h^{-1}, and TMP 5 × 10^4 Pa. Working with these conditions, on the basis of cell density increase and glucose concentration decay, the most convenient starting time for ultrafiltration was around 7 h from the onset of the fermentation. A concentration of lactic acid of 8–10 g l^{-1} in the permeate was obtained. The total amount of lactic acid produced in about 20 h by continuous fermentation ($g_{fermentor}$ + $g_{permeate}$) was higher by 40% than that from batch fermentation (Figure 3.2b).

It is expected that the performance of the continuous membrane fermentor can be improved by integrating it into a continuous downstream process that selectively separates the lactic acid and recycles the substrate and nutrients, permeated through the UF membrane, back to the fermentor solution.

It is generally observed that membrane fouling during microfiltration or ultrafiltration of fermentation broth causes permeate flux decay, which decreases the performance of the process and increases costs. Several studies have been carried out aimed at limiting the effects of fouling. Schlosser (1993b) has shown that it is possible to reduce the negative effects of fouling during microfiltration of fermentation broths by operating in back-flush microfiltration mode. In microfiltration of lactic acid broth through ceramic membranes, periodic short back-flushing of permeate greatly increases the permeate flux, which is highest at a back-flush frequency of about 1 min^{-1}; at a transmembrane pressure of 110 kPa the flux was four times higher. The efficiency of back-flushing strongly depends on the fouling mechanisms, and on the kind of fermentation broth that is treated.

Visacky et al. (1996) at LIKO Research (Slovakia) Institute studied continuous fermentation for production of carboxylic acids such as citric acid, acetic acid, kojic acid and tartaric acid. Citric acid was produced by submerged fermentation of *Aspergillus niger*. The fermentation broth was ultrafiltered through industrial type plate-and-frame modules UFZ400 (Liko design). The permeate recovered from the ultrafiltration units was concentrated through spiral-wound nanofiltration modules. High shear in the circulating pump may cause break up of the *Mycelia hyphae*, which may then have a negative effect on citric acid production. For this reason, a pretreatment step is necessary before ultrafiltration. On the basis of results of various projects, an integrated process scheme for citric acid production has been suggested (Figure 3.3).

Recovery of Carboxylic Acids by Integrated Membrane Processes

Carboxylic acids obtained in aqueous solutions by continuous fermentation have to be separated from excess reagents, substrates, nutrients, etc., from impurities introduced through feed, and from by-products. Separation is usually followed by final purification, concentration and crystallization.

The most common membrane separation processes used for these steps will be discussed in the following sections. These membrane operations can be used in integrated systems as indicated in Figure 3.4.

Electrodialysis The application of electrodialysis in separation and concentration of organic acids is well documented in the literature (Ohleyer et al., 1985; Taniguchi et al., 1987; Bibal et al., 1991). It offers possible solutions for the separation and concentration of ionic substances, such as organic acids, in the food and pharmaceutical industries.

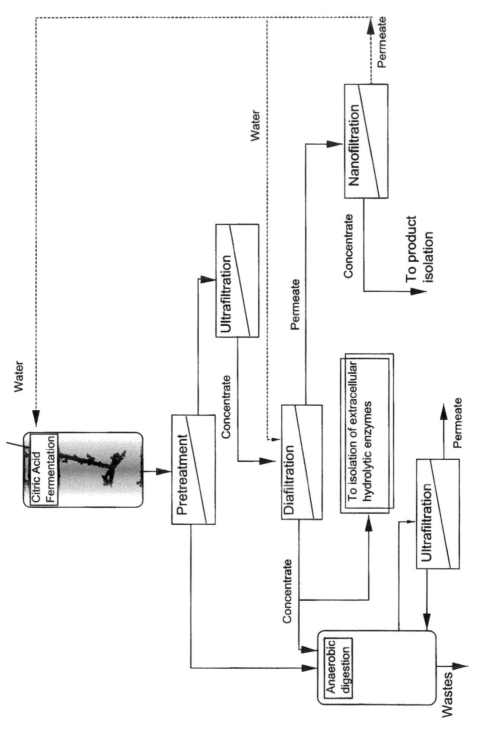

Figure 3.3 Continuous fermentation of citric acid in an integrated process (after Visacky et al., 1996).

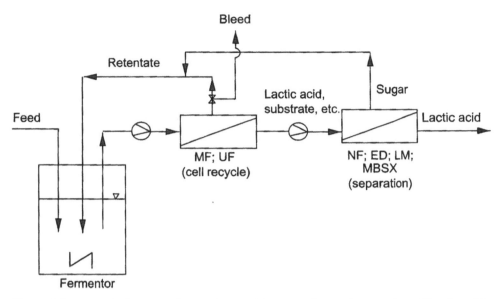

Figure 3.4 Integrated process for continuous bioconversion and separation of lactic acid.

In fermentation technology, it has been used to isolate inhibitive metabolites to allow continuous fermentation, especially for lactic acid. Various combinations of membrane arrangements have been suggested. The production rate of systems using this technique can also be increased compared with traditional fermentation methods (Boyaval and Corre, 1987).

Lactic acid is produced by *Lactobacillus* in fermentation broths kept at pH 6.0 by addition of carbonates (potassium and calcium). It is then present in the salt form, and needs to be separated and converted to the acid form. Electromembrane techniques enable both conversion and recovery to be performed. (Citric acid is produced by *Aspergillus* at pH values down to 3. The product then appears in the native (acidic) form at a concentration up to 130 g l⁻¹. Thus, the technique for recovery of citric acid should enable its separation from the fermentor constituents by ultrafiltration and other membrane techniques.)

Electrodialysis involves transport of organic ions. The organic ions can pose problems at an anion exchange membrane. As the transport numbers of mobile ions in the membranes are different from those in the solution, concentration polarization takes place at the ion exchange membrane–solution interface. In addition, membrane fouling might occur. The phenomenon is particularly severe at anion exchange membranes with large anions of organic origin. Owing to concentration polarization there exists a limiting current level at which interfacial concentration becomes zero. The boundary layer thickness is influenced by fluid dynamics, hence influencing the performance. Murdia et al. (1986) investigated the influence of hydrodynamics and kinetics on the electrodialytic process. Sodium lactate was the model substance. The hydrodynamics in the stack and the critical flow conditions were found to be dependent on temperature and on the dimension of the channel. The mass transfer coefficient, expressed as a modified Nusselt number, is influenced weakly by the Reynolds number in the laminar region, but strongly in the turbulent region, while the influence of the Schmidt number is approximately similar in both regions. The flow through an electrodialysis stack becomes complex owing to the presence of spacers that are used to increase the turbulence. The typical configuration of spacers changes the flow dynamics and causes a change in the value of the Reynolds number.

An electrochemical method has been proposed to determine the critical flow conditions. In this method, the potential drop (proportional to the electrical resistance) across the channel containing the deionized solution is recorded as a function of flow velocity. To investigate the effects of channel dimensions, the experiments are done with the channel having one, two or three spacers. Platinum electrodes are used to measure the potential drop across a diluate channel formed by an anion exchange and a cation exchange membrane.

At constant current, the potential drop gradually decreases and reaches a constant value when the flow velocity is gradually increased and ultimately becomes highly turbulent. It is ensured that the concentration of the solution remains unchanged. Pressure gauges are also installed to record the pressure drop across the diluate channel as a function of flow velocity, to verify the results obtained by electrochemical methods.

An electrodialysis apparatus can be combined directly with a fermentor. In this way it is possible to remove the lactic acid selectively as it is produced. Disadvantages of this operation are partial loss of lactic acid by diffusion through the cation exchange membrane into the acid solution of the anode recirculation, and increase in the electrical resistance as a consequence of the adherence of bacteria to the membrane. This latter disadvantage can be solved by microfiltration of the fermentation broth, as suggested by van Nispen and Jonker (1991). They studied a process for fermentative production of lactic acid by continuous fermentation in a reaction vessel, with the fermented reaction mixture being ultrafiltered. The retentate is recycled to the reaction vessel and the permeate is concentrated by reverse osmosis and subsequently subjected to electrodialysis using bipolar membranes. Electrodialysis with bipolar membranes has the advantage that no external protons have to be added because of water splitting at the anionic–cationic interface of the bipolar membrane (Strathmann, 1986).

An integrated approach using electrodialysis and bipolar membranes is elegant but expensive. Nevertheless, a combination of separately operated units forms an interesting option from an economic and environmental point of view by avoiding the generation of salty streams.

Diffusion dialysis Diffusional permeation of a penetrant driven by its concentration gradient is economical; however, this is a technique with a rather low transport rate. An increase in the transport rate is possible if the permeant reacts with the functional groups of the polymer used for membrane preparation. This technique is known as 'facilitated diffusion'. The effect of acceleration depends very much on the nature of the permeant and on the chemistry of the membrane.

Narebska (1992) studied a new integrated membrane system for recovery of lactic acid consisting of diffusion dialysis (Dd) and electrodialysis (ED) modules operating successively. The novelty is in using diffusion dialysis instead of ultrafiltration. By diffusion dialysis the acid is continuously isolated from the broth to the pH-controlled alkaline solution in the receiving side of a dialyzer (external neutralization) and is concentrated as sodium lactate to about 0.5–0.6 mol l^{-1}. No energy is consumed at this stage, except for pumping. In the subsequent step, sodium lactate is converted into the acid by electrodialysis. Two electrodialyzers, one with monopolar membranes (ED-MP) and one with bipolar (ED-BP) were examined, and for both interrelations between the feed and product concentration, current density and energy consumption per unit of the acid were established. With Dd-ED, lactic acid was isolated as an aqueous solution of concentration 1 mol l^{-1} or higher per 3.5–4.5 kWh kg^{-1}.

The advantage of diffusion dialysis is the continuous isolation of carboxylic acids from the fermentor as they are produced. With removal of the acid it is not necessary to neutralize the fermentation broth. Membranes for diffusion dialysis are of gel type, so they are rather impermeable to colloids and the permeation of any inorganic salts will be much depressed against the acid. Separation factors of the acid against salt of about 20 with Neosepta membranes and up to 30 with Selemion membranes have been reported.

The efficiency of diffusion dialysis is specially high at low concentration of the acid (0.01 mol l⁻¹).

Reverse osmosis (RO) and nanofiltration (NF)   Reverse osmosis using thin-film composite membranes has been applied to concentrate dilute lactic acid solutions (Schlicher and Cheryan, 1990). The lactic acid rejection of the composite and cellulose acetate RO membranes was found to be strongly dependent on pH. This was explained by assuming that undissociated lactic acid permeates freely with the water through the membrane while the dissociated form is rejected.

Lactic acid fermentation and simultaneous removal of lactic acid in RO membrane reactors has also been studied (Hanemaaijer et al., 1988). This membrane process has a large influence on the economics of the process, and therefore a better understanding of the mass transfer process of lactic acid through RO membranes is necessary.

Timmer et al. (1993) developed a mass transfer model to describe the separation of lactic acid with the help of RO and NF membranes. They used RO membranes (DDS HR95 and DDS CH995) and NF membranes (DDS HC50 and DDS CA960) from NIRO Atomizer (Apeldvoron, Netherlands). The CA-type membranes were made of cellulose acetate and the HC- and HR-types were thin-film composite membranes composed of a polyamide separation layer on a polysulfone support. The experiments were carried out in a DDS Lab-20 unit in which the four membranes were installed in series. Each membrane was 36×10^{-3} m long. Experiments were performed in batch recirculation mode, which means that both the permeate and concentrate were recycled back to the feed vessel. The circulation velocities applied were 7.4 to 10 l min⁻¹ in order to test whether concentration polarization was occurring.

The results showed that for both circulation velocities the flux through an HC50 membrane was directly proportional to the transmembrane pressure. This indicates that concentration polarization phenomena are of minor importance and the mass transfer resistance in the boundary layer on the upstream side of the membrane is of minor importance to the overall mass transfer resistance across the membrane. The same observations were made for the CA960 and CA995.

These observations are in agreement with the results of Schlicher and Cheryan (1990), who developed a model based on the extended Nernst–Planck equation for describing the mass transfer of lactic acid through RO and NF membranes. The model can be used to determine mass transfer characteristics of different RO and NF membranes with the assumption that undissociated and dissociated lactic acid are separate components. Furthermore, the model was applied to quantify lactic acid rejection and fouling of RO and NF membranes and to determine the major fouling mechanisms on laboratory-scale and pilot-plant NF and RO experiments with fermentation broth. Especially at high fluxes, the prediction of lactic acid rejection was quite good.

Fouling of the membrane could be quantified in terms of three resistances: a membrane resistance, an initial fouling resistance, and a time-dependent fouling resistance. Equations for the initial fouling resistances were developed and time-dependent fouling could be described either by a colloidal fouling model (ultrafiltered fermentation broth) or a gel layer model (fermentation broth).

Evaluation of the three resistances by simulation of continuous and batch concentration experiments showed that during NF of an ultrafiltered fermentation broth the initial fouling resistance, resulting from concentration polarization effects, was the predominant resistance.

For a fermentation broth the time-dependent fouling becomes more important than the initial fouling resistance. Protein fouling is the main cause of the time-dependent fouling. Therefore, it is recommended to remove proteins by ultrafiltration before NF.

Lactic acid rejection of composite and cellulose acetate membranes was found to be strongly dependent on pH (Schlicher and Cheryan, 1990; Timmer et al., 1993). This was explained by a difference in ion rejection of the membrane for dissociated and undissociated lactic acid.

NF membranes in general show lower rejection for lactic acid than RO membranes. For the selective separation of lactic acid from a fermentation broth, this is an advantage compared to RO.

For industrial applications, a predictive model for the lactic acid rejection of NF and RO membranes that can be used for process development or optimization is needed. The model described by Timmer et al. (1993) can serve as a starting point for describing and predicting lactic acid rejection of NF and RO membranes using fermentation broths.

In addition to description of the lactic acid rejection, the development of the flux must also be taken into account. Models need to be developed that take account of the initial fouling phenomena and describe a steady-state situation in which long-term fouling occurs.

Pertraction Pertraction by liquid membranes (LM), supported liquid membranes (P-SLM) and membrane-based solvent extraction (MBSX) has been widely used for the separation of organic acids.

The systems in which the pertraction occurs are called contactors, named on the basis of membrane nature and configuration: pertraction by LM in a bulk contactor; pertraction by MBSX in hollow-fibre contactors, and so on (see Chapter 1).

Carboxylic acid separation is also obtained through enzyme-mediated reactions in liquid membranes (Rethwisch et al., 1990). The literature describes many potential advantages of LMs as an industrial separation method (Danesi, 1986; Ho, 1990; Reisinger et al., 1990; Way et al., 1982; Coelhoso et al., 1995, 1996). These advantages include high selectivities, high fluxes, use of an external driving force to concentrate the product during separation, low capital and operation costs, and compact installation. In spite of these apparent advantages, very few industrial applications have been reported so far (Way et al., 1982; Dworzak and Naser, 1987). Some drawbacks were shown to hinder implementation, the most important being membrane instability in SLM, complexity of operation and swelling in ELM.

Schlosser (1993a) studied pertraction through liquid membranes and solvent extraction with membranes for the recovery of organic acids from fermentation broths.

Screening of carriers for pertraction of organic acids based on equilibrium data and on the toxicity of carriers in the fermentation production of acids have been carried out. The solvent toxicity is strongly influenced by the nature of the microorganism used (Martak et al., 1994a, 1994b).

A hollow-fibre-in-tube pertractor (Schlosser, 1993b) has also been developed. The feed (0.1 mol l^{-1} NaOH; $V_r = 0.37$ cm^3 min^{-1}) and the stripping solution (lactic acid 46.8 g l^{-1}, pH 1.9; $V_f = 1.35$ cm^3 min^{-1}) were flowing in polypropylene hollow fibres. The lactate concentration in the loaded stripping solution was 5.7 g l^{-1} and pH$_R$ was 12.4). Using

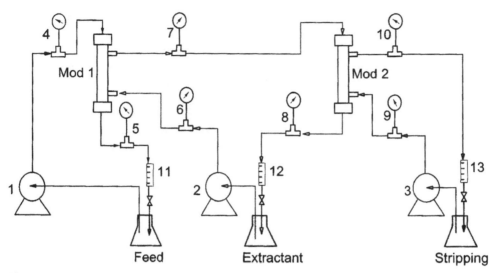

Figure 3.5 Membrane-based solvent extraction in hollow-fibre membrane contactors; items 1 to 3 are pumps; 4 to 10 are pressure ganges; 11 to 13 are flow meters.

these conditions, a productivity per unit volume of the pertractor of 230 mol $h^{-1}m^{-3}$ (20.9 kg $h^{-1}m^{-3}$) was obtained.

Pertraction of lactic acid by membrane-based solvent extraction was also investigated by Giorno et al. (1996). This process employs an organic carrier that picks up the anions of lactic acid from the feed solution to an extraction module and releases them during contact with the stripping solution in a re-extraction module. Contact between feed, organic solution and stripping is established by the porous membranes, which are preferentially wetted by one of the phases. A controlled pressure drop is then necessary to stabilize the separated-multiphase system.

Experiments were performed using hollow-fibre-in-tube modules (1, 2 in Figure 3.5) which consisted of 4 or 6 fibres glued by epoxy resin. The solutions were recirculated in a countercurrent flow regime. The effect of carrier on the separation efficiency has been investigated. The highest performance has been obtained with secondary amines, according to Reisinger and Marr (1992). The optimum operating conditions, such as type of carrier and carrier concentration, lactic acid concentration in the feed, temperature and pH, have been identified (Giorno et al., 1997).

Three different carrier concentrations have been evaluated: 5, 10 and 15% w/v of LA-2 in n-heptane. Experiments were carried out using 20 g l^{-1} of D,L-lactic acid in distilled water at pH 5.0 as feed solution. The behaviour of the transport rate as a function of carrier concentration gives an optimum value. The overall effect is due to the compromise between the improvement of the transport due to the presence of there being more carrier available to combine and transport more solute and the negative effect of viscosity on the diffusion coefficient, as predicted by the Stokes–Einstein equation.

The effect of lactic acid concentration on the separation efficiency has been investigated using 20 and 40 g l^{-1} of D,L-lactic acid in distilled water at pH ~5. The extractant phase consisted of 10% LA-2 in n-heptane. The extraction efficiency increases at higher lactic acid concentration. The amount of lactic acid in the fermentation broth depends on the type of microorganism used. Bacteria, such as *Lactobacillus bulgaricus*, produce about 20 g l^{-1}, while fungi, such as *Aspergillus niger*, produce over 40 g l^{-1} of lactic acid.

A disadvantage of using fungi is the decrease of permeate flux due to their capacity to grow on the membrane.

The influence of temperature on the extraction efficiency has been evaluated with 20 g l^{-1} of D,L-lactic acid in distilled water, pH ~5, extractant 10% LA-2 in *n*-heptane. The transport rate as a function of temperature initially increases, owing to the increase of diffusion coefficient with the temperature. Subsequently, the transport rate decreases with increasing temperature, owing to the destabilization of the carrier–solute complex.

The membrane-based solvent extraction system showed stability for over 5 days of continuous operation. In terms of degree of separation, it can be competitive with other techniques such as liquid membrane extraction (Martak et al., 1994a) but is less efficient than a liquid emulsion membrane (Reisenger and Marr, 1992); on the other hand, it is much more stable.

3.2.3   *Production and Separation of Antibiotics*

Antibiotics are low molecular mass organic compounds produced as secondary metabolites during the fermentation of filamentous organisms. The first step in the recovery of antibiotics is the separation of the microorganisms and other solids in the fermentation broth. The clarified broth is then separated by extraction or by selective chromatographic methods.

Membrane technologies have been applied in the pharmaceutical industry to replace some of the traditional techniques. Gravott and Molnar (1986) proposed the use of ultrafiltration to separate antibiotics. Using a recirculation rate of 8000 l min^{-1} at an operating pressure of 0.4 MPa, a permeate flux of about 80 l min^{-1} was reached.

The use of membrane processes combined with HPLC for the separation of antibiotics was suggested by Sartori et al. (1985). Cephalosporin-C was used as an antibiotic model. The fermentation broth was initially clarified through a crossflow microfiltration system, to separate cells. The permeate was then processed by ultrafiltration to eliminate high molecular mass proteins and polysaccharides. The ultrafiltration permeate was concentrated by reverse osmosis and then purified by HPLC. Antibiotic retention by the membrane was very high, greater than 99%.

Production of penicillin G is far greater than that of any other antibiotics in terms of total capital invested in plant facilities, return on capital investment, and total volume of antibiotics produced. Generally, the precursor phenylacetic acid (PAA) is often added to the fermentation medium to promote the production of penicillin G during fermentation. The expensive precursor PAA must be separated and recovered from the fermentation broth because of its toxicity. However, the separation is not easy because the properties of penicillin G and PAA are quite similar. Amberlite LA-2 was used as organic carrier, dissolved in 1-decanol and supported in a microporous polypropylene membrane. A maximum separation factor of 1.8 was achieved under liquid membrane mass transfer control (Lee et al., 1993, 1994). SLMs have been used to separate penicillin (Marchese et al., 1989) as well as emulsion liquid membranes (Hano et al., 1990; Lee and Lee, 1992; Lee at al., 1997).

To test the efficiency of separation processes, simulated aqueous solutions of penicillin are in general used instead of real fermentation broth. These solutions are prepared by dissolving penicillin in a citrate buffer solution of 0.5 mol l^{-1}. An organic carrier (amine, such as amberlite LA-2) with a surfactant (non-ionic polyamine, such as ECA 4360J) diluted in organic solvent (heptane, kerosene) can be used as the organic solution. Buffer solution at basic pH (such as sodium carbonate) can be used as the receiving phase. A homogenizer or similar apparatus is used to create an oil-in-water emulsion. In general the penicillin acid anion and hydrogen ion are co-transported from the feed to the receiving phase by the carrier, which reacts with them at the feed/organic phase interface and releases them to the interface with the high-pH receiving phase.

The parameters that influence the performance of the processes in emulsion liquid membranes are the stirring speed, the oil/water ratio, the concentration of penicillin in the feed solution, the type of carrier and carrier concentration, and the concentration and pH of the receiving phase. These parameters influence the efficiency of the process because they affect the transport across the interfaces between the different phases.

3.3 Integrated Membrane Systems in the Processing of Natural Products

Membrane processes can be used for the treatment of various juices and musts. They have been used for the concentration of must using reverse osmosis after pretreatment by ultrafiltration; for ultrafiltration of must to minimize the use of SO_2; for controlling of concentration of polyphenols in musts and wines; and for clarification of wine after fermentation. An example of integrated membrane processes suggested for wine production is shown in Figure 1.19 (Chapter 1) (Drioli and Molinari, 1990).

The use of microfiltration has several advantages, such as reduced use of SO_2 owing to low concentration of microbes; higher stabilization of tartrate owing to the removal of colloids that delay its precipitation; higher quality and improved taste of wine; better colour and body of red wines; and no undesirable change in wine composition.

Ultrafiltration and enzyme membrane reactors can also be used in the process of winemaking. Ultrafiltration allows removal of microorganisms that could produce undesired biotransformations. After sterilization, the desired microorganism for carrying on the fermentation can be inoculated.

During maturation of wine, after fast fermentation of sugars into alcohols, a secondary fermentation occurs that converts malic acid into lactic acid. The possibility of controlling these fermentations allows a high quality product with improved organolectic properties to be obtained. Membrane bioreactors using immobilized *Leuconostoc oenos* have been used to control malolactic fermentation of white wine. Bacteria were immobilized by crossflow microfiltration in polypropylene membranes (pore diameter 0.2 µm). To obtain good performance it is better to activate the bacteria and immobilize them when they are in the exponential growth phase. In Figure 3.6 it is shown that malolactic fermentations carried out with free and immobilized cells of *Leuconostoc oenos* have almost the same behaviour. The advantage of membrane bioreactors lies in the possibility of controlling fermentation by removing the permeate when the desired level of malic acid conversion is reached. These levels can be controlled by measuring the pH of wine, which increases slightly as deacidification occurs (Figure 3.7).

Membrane technology has also been used successfully in fruit juice processing for depectinization, clarification, stabilization, and concentration (Cheryan, 1984; Rao et al., 1987; Shen et al., 1987). Integrated membrane operations are used to improve organolectic properties of concentrated juices and reduce their costs.

Pectins are responsible for the turbidity and high viscosity of fruit juices. They are linear polymers composed of α-1,4-linked D-galacturonic acid units, characterized to a certain extend by methylation of their carboxylic group (Fogarty and Kelly, 1983; Borrego

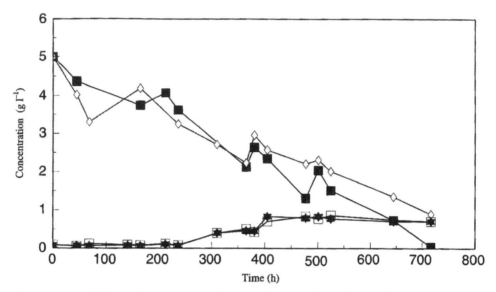

Figure 3.6 Malolactic fermentation behaviour of free and immobilized bacteria; T = 28 °C, pH 3.45. L-Malic: ■, batch; ◇, MBR. L-Lactic; ✱, batch; □, MBR.

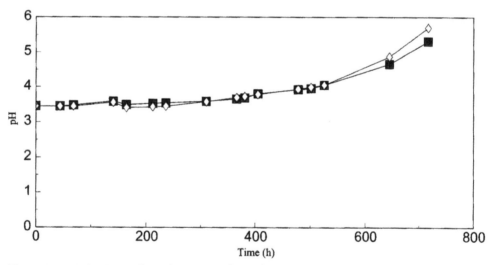

Figure 3.7 Behaviour of pH during malolactic fermentation carried out in batch (◇) or in membrane bioreactor (■).

et al., 1989). Along the galacturonic acid chain, sugar units are present as side groups. The viscosity of fruit juices increases as the concentration of high molecular mass pectins increases. The hydrolysis and fractionation of pectins allows control of the viscosity and turbidity of the juices. The use of integrated membrane processes to produce concentrated apple juice has been the goal of a multinational research project sponsored by the European Union (AIR3-CT94-1931).

A flowsheet of the proposed overall process is shown in Figure 3.8. The refined raw juice is treated in an enzyme membrane reactor that combines the hydrolysis of pectins and the separation of depectinized juice by ultrafiltration. The clarified juice is

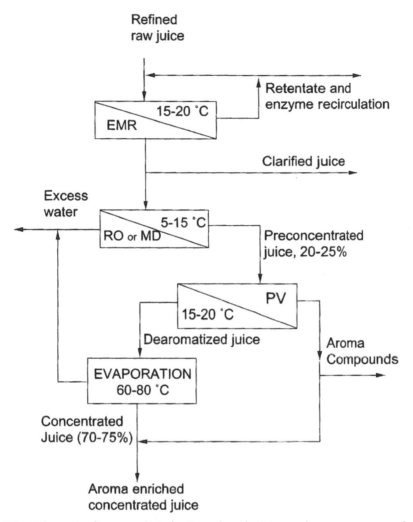

Figure 3.8 Schematic diagram of production of apple juice and aroma compounds by integrated membrane processes. EMR, enzymatic membrane reactor; RO, reverse osmosis; PV, pervaporation.

then preconcentrated by reverse osmosis or membrane distillation. In this way, the excess water can be reused and the preconcentrated juice can be concentrated again by pervaporation. This operation allows recovery of the aroma compounds and preserves them from the evaporation process (60–80 °C). After evaporation, the concentrated juice is enriched with the aroma compounds.

The parameter that most affects the enzyme membrane reactor performance in terms of permeate flux is membrane fouling. In addition to fluid dynamic conditions, fouling can be controlled in these systems using the enzyme immobilized on or in the membrane (Szaniawski and Spencer, 1997; Giorno et al., 1998). In this way, cake layer formation is prevented by the hydrolytic action of the enzyme as pectins deposit on the membrane surface. As shown in Figure 3.9, the normalized permeate flux of apple juice treated with immobilized enzyme is higher than that of the juice treated with free enzyme. The optimal amount of protein to immobilize on the membrane is a parameter that has to be

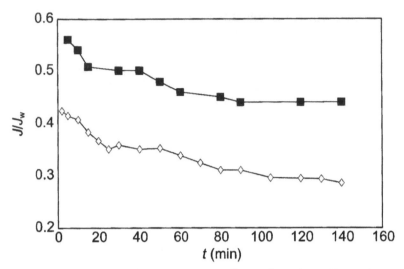

Figure 3.9 Permeate flux J relative to pure water flux J_w through U002 spiral-wound module of juice treated with free ($\diamond$) and immobilized rapidase ($\blacksquare$): $Q_{ax} = 800\ \mathrm{l\ h^{-1}}$; $T = 25\ ^{\circ}C$; transmembrane pressure = 0.7 kPa.

optimized case by case. Crossflow microfiltration of low-pectin apple juice was also studied by Alvarez et al. (1996) using 0.2 μm inorganic membranes (TechSep Carbosep, France), having a minimal pore size of 0.2 μm. High and stable permeate fluxes were obtained. Membrane fouling was not an important drawback and pectins were almost completely retained by the membranes.

The treatment of milk and wheys is another important example of integrated membrane operations. The principal membrane processes used in the dairy industry are reverse osmosis (De Boer and Nooy, 1980) to concentrate milk and whey; electrodialysis to demineralize whey; ultrafiltration to fractionate and concentrate proteins; microfiltration to remove microorganisms and fats; enzyme membrane reactors to hydrolyze proteins and lactose (Coca et al., 1992). Animal milk contains allergenic high molecular mass proteins, among which the most allergenic is β-lactoglobulin. The hydrolysis of proteins into peptides of low molecular mass allows the production of nutritious, nonallergenic milk, useful as baby food. Trypsin and chymotrypsin enzymes can be used in membrane reactors for this purpose.

Whey is the liquid by-product of cheese and casein manufacturing. It contains 50–65% of the initial milk solids, most of the lactose, 20% of the protein, and most of the vitamins and minerals. Cheese whey has been used for the recovery of selected whey products. Typical whey constituents are lactose (44 g l⁻¹), proteins (~7 g l⁻¹), ash (5 g l⁻¹) and fats (0.2 g l⁻¹). Proteins and lactose are the components to be recovered. Whey proteins account for only 0.7%, but proteins such as α-lactoalbumin, β-lactoglobulin and immunoglobulins have nutritional and pharmacological activities.

Lactose and ash can be separated from proteins by ultrafiltration (Coca et al., 1992). Ultrafiltration of raw whey through flat polysulfone membranes with 20 kDa cut-off concentrated the whey proteins up to 56 wt%.

Reverse osmosis removes only water, increasing the total solid concentration of the whey. It can be used to concentrate raw whey or lactose and salts from ultrafiltration permeate.

Electrodialysis is used to demineralize whey, which is necessary if whey is used as a foodstuff. Electrodialysis reduces the salt content by up to 60–70%.

3.4 Conclusions

Continuous membrane fermentors in integrated systems for the production and separation of bioactive compounds represent a promising new technology. It has largely been studied at laboratory level, showing the potential advantages in terms of efficiency, sensitivity towards biological materials, feasibility of scale-up, positive environmental impact, and so on. It needs to be explored and verified at industrial level.

The high added value of low molecular mass products that can be processed with this technology and the social impact of biological compounds obtained by mild treatments promote the growth of research and demonstration projects aimed at the deeper exploitation of these potentials.

3.5 References

ALVAREZ, V., ANDRES, L.J., RIERA, F.A. and ALVAREZ, R., 1996, Microfiltration of apple juice using inroganic membranes: process optimization and juice stability, *Can. J. Chem. Eng.*, **74**, 156–162.

ARNOLD, M.H.M., 1975, *Acidulants for Food and Beverages*, p. 58, Food Trade Press, London.

BIBAL, B., VASSIER, Y., GOMA, G. and PAREILLEUX, A., 1991, High concentration cultivation of *Lactococcus cremoris* in a cell recycle reactor, *Biotechnol. Bioeng.*, **37**, 746.

BORGLUM, G.B. and MARSHALL, J.J., 1984, *Appl. Biochem. Biotechnol.*, **9**, 117–130.

BORREGO, F., TARI, M., MANOJIN, A. and IBORRA, J.C., 1989, *Appl. Biochem. Biotechnol.*, **22**, 129–140.

BOYAVAL, P. and CORRE, C., 1987, Continuous lactic acid fermentation with concentrated product recovery by ultrafiltration and electrodialysis, *Biotechnol. Lett.*, **9**, 207–212.

BUCHTA, K., 1974, *Chem. Z.*, **98**, 532–538.

CHERYAN, M., 1984, *Ultrafiltration Handbook*, Technomic Publishing, Lancaster, PA.

CHERYAN, M. and MEHAIA, M.A., 1985, Continuous fermentation with membrane bioreactors, *4th Int. Congress Eng. Food*, Alberta, Canada.

COCA, J., ALVAREZ, R. and ALVAREZ, S.R., 1992, Bioseparators with membranes, presented at *EC-Membrane Workshop, Brazil '92*, Rio de Janeiro, pp. 212–27.

COELHOSO, I.M., MURA, T.F., CRESPO, J.P.S.G. and CARRONDO, M.J.T., 1995, Transport mechanisms in liquid membranes with ion exchange carriers, *J. Membr. Sci.*, **108**, 221–224.

COELHOSO, I.M., CRESPO, J.P.S.G. and CARRONDO, M.J.T., 1996, Modeling of ion-pairing extraction with quaternary amines, *Sep. Sci. Technol.*, **31**(4), 491–511.

DANESI, P.R., 1986, Support liquid membranes in 1986. New technologies or scientific curiosity. *Proc. Int. Solvent Extraction Conf.*, Munchen, Germany, Vol. I, pp. 527–536.

DE BOER, R. and NOOY, P.F.C., 1980, Concentration of raw whole milk by reverse osmosis and its influence on fat globules, *Desalination*, **35**, 201.

DRIOLI, E., 1986, Membrane processes in the separation, purification, and concentration of bioactive compounds from fermentation broths, *Am. Chem. Soc.*, **5**, 52–66.

DRIOLI, E. and MOLINARI, R., 1990, Membrane processing of wine and alcoholic beverages, *Chemistry Today*, 47–55.

DRIOLI, E., IORIO, G., DE ROSA, M., GAMBACORTA, A. and NICOLAUS, B., 1982, High temperature immobilized-cell ultrafiltration reactors, *J. Membr. Sci.*, **11**, 365.

DWORZAK, W.R. and NASER, A.J., 1987, Pilot-scale evaluation of supported liquid membrane extraction, *Sep. Sci. Technol.*, **22**(2–3), 677–689.

EIKMEIER, H. and REHM, H.J., 1984, *Appl. Microbiol. Biotechnol.*, **20**, 365–370.

FANE, A.G., FELL, C.J.D. and WALTERS, A.G., 1981, The relationship between membrane surface pore characteristics and flux for ultrafiltration membranes, *J. Membr. Sci.*, **9**, 245–262.

FOGARTY, W.M. and KELLY, C.T., 1983, Pectic enzymes, in FOGARTY, W.M. (ed.) *Microbial Enzymes and Biotechnology*, pp. 131–182, Applied Sciences Publishers, London.

GIORNO, L., SPICKA, P. and DRIOLI, E., 1996, Downstream processing of lactic acid by membrane-based solvent extraction, *Sep. Sci. Technol.*, **31**(16), 2159–2169.

GIORNO, L., RAIMONDI, G. and DRIOLI, E., 1997, Integrated membrane operations for recovery of lactic acid from fermentation broth, *Proc. First Eur. Congr. Chemical Engineering — ECCE1*, Florence, pp. 1241–1244.

GIORNO, L., DONATO, L., TODISCO, S. and DRIOLI, E., 1998, Study of fouling phenomena in apple juice clarification by enzyme membrane reactor, *Sep. Sci. and Technol.*, **33**(5), 739–756.

GRAVOTT, D.P. and MOLNAR, T.E., 1986, Recovery of an extracellular antibiotic by ultrafiltration, in MCGREGOR, W.C. (ed.) *Membrane Separation in Biotechnology*, Marcel Decker, New York, pp. 115–27.

HANEMAAIJER, W.I., TIMMER, J.M.K. and JEURNINK, T.J.M., 1988, Continue produkte van melkzuur in membraanreactoren, *Voedingmiddelentechnologie*, **21**(9), 17.

HANG, Y.D. and WOODAMS, E.E., 1984, *Biotechnol. Lett.*, **6**, 763–764.

HANO, T., OHTAKE, T., MATSUMOTO, M., OGAWA, S. and HORI, F., 1990, Extraction of penicillin with liquid surfactant membrane, *J. Chem. Eng. Jpn.*, **23**(6), 772.

HEFNER, D. and GIORNO, L., 1992, Internal report, Sepracor Inc., USA.

HIDDINK, I., DE BOER, R. and NOOY, P.F.C., 1980, Reverse osmosis of dairy liquids, *J. Dairy Sci.*, **63**, 204.

HO, W.S., 1990, Emulsion liquid membranes: a review. *Proc. Int. Congr. Membranes and Membrane Processes*, Chicago, Vol. I, pp. 692–694.

HOLTEN, C.H., 1971, *Lactic Acid*, pp. 192–231, Verlag Chemie, Weinheim.

HOSSAIN, M., BROOKS, J.D. and MADDOX, I.S., 1983, *NZ J. Dairy Sci. Technol.*, **18**, 161–168.

IWASAKI, K., NAKAJIMA, M. and SASAHARA, H., 1993, Rapid continuous lactic acid fermentation by immobilized lactic acid bacteria for soy sauce production, *Process Biochem.*, **28**, 39–45.

LEE, C.-J., YEH, H.-J., YANG, W.Y. and KAN, C.-R., 1993, Extractive separation of penicillin G by facilitated transport via carrier supported liquid membranes, *Biotechnol. Bioeng.*, **42**, 527–534.

LEE, C.-J., YEH, H.-J., YANG, W.Y. and KAN, C.-R., 1994, Separation of penicillin G from phenylacetic acid in a supported liquid membrane, *Biotechnol. Bioeng.*, **43**, 309–313.

LEE, S.C. and LEE, W.K., 1992, Extraction of penicillin G from simulated media by an emulsion liquid membrane process, *J. Chem. Technol. Biotechnol.*, **53**, 251.

LEE, S.C., LEE, K.H., HYUN, G.H. and LEE, W.K., 1997, Continuous extraction of penicillin G by an emulsion liquid membrane in a countercurrent extraction column, *J. Membr. Sci.*, **124**, 43–51.

LEHNINGER, A.L., 1987, *Biochimica*, pp. 375–397, Zanichelli, Milan.

MARCHESE, J., LÒPEZ, J.L. and QUINN, J.A., 1989, Facilitated transport of benzyl penicillin through immobilized liquid membrane, *J. Chem. Technol. Biotechnol.*, **46**, 149.

MARGARITIS, A. and WILKE, C., 1978, The rotofermentor. I. Description of the apparatus, power requirements, and mass transfer characteristics, *Biotechnol. Bioeng.*, **22**, 709–726.

MARTAK, J., SCHLOSSER, S. and STOPKA, J., 1994a, Pertraction of lactic acid through liquid membranes, *Proc. 7th Symp. Synthetic Membranes in Science and Industry*, Tubingen.

MARTAK, J., ROSENBERG, M., SCHLOSSER, S. and KRISTOFIKOVA, 1994b, Organic solvent toxicity during fungal fermentation of lactic acid, *IUMS Congress '94*, Prague, Poster MS112–323.

MATSUURA, T., 1973, *J. Appl. Polym. Sci.*, **17**, 3663–3668.

MEHAIA, M.A. and CHERYAN, M., 1987, Production of lactic acid from sweet whey permeate concentrates, *Process Biochemistry*, **22**, 185–188.

MILSOM, P.E., 1986, Organic acids by fermentation, especially citric acid, in KING, R.D. and CHEETHAM, P.S.J. (eds), *Food Biotechnology 1*, pp. 273–307, Elsevier Applied Science, London.

MORISI, F., MOSTI, R. and BETTONTE, M., 1976, Snam Progetti SpA, U.K. Patent 1,426,137.

MOUEDDEB, H., SANCHEZ, J., BARDOT, C. and FICK, M., 1996, Membrane bioreactor for lactic acid production, *J. Membr. Sci.*, **114**, 59–71.

MURDIA, L.K., REHMANN, D. and BAUER, W., 1986, Influence of hydrodynamics and kinetics on the performance of lactic acid concentration by electrodialysis, in SPIESS, W.E.L. and SCHUBERT, H. (eds), *Engineering and Food*, Vol. 3, pp. 253–262, Elsevier Applied Science, London.

NAKAO, S., NOMURA, T. and KIMURA, S., 1979, Characteristics of macromolecular gel layer formed on ultrafiltration tubular membrane, *AIChEJ*, **25**, 615–622.

NAREBSKA, A., 1992, *Second Annual Report of Copernicus Project*, ERB-CIPA CT92 3018.

NILSON, P.E., 1986, Organic acids by fermentation, specially citric acid, in *Food Biotechnology*, p. 289.

OHARA, H., HIYAMA, K. and YOSHIDA, T., 1992, Kinetics of growth and lactic acid production in continuous and batch culture, *Appl. Microbiol. Biotechnol.*, **37**, 544–548.

OHARA, H., HIYAMA, K. and YOSHIDA, T., 1993, Lactic acid production by a filter-type reactor, *J. Ferment. Bioeng.*, **76**(1), 73–75.

OHLEYER, E., WILKE, C.R. and BLANCH, H.W., 1985, Continuous production of lactic acid from glucose and lactose in a cell recycle reactor, *Appl. Biochem. Biotechnol.*, **11**, 457–463.

RAO, M.A., ACREE, T.E., COOLEY, M.J. and ENNIS, R.W., 1987, Clarification of apple juice by hollow fibre ultrafiltration: fluxes and retention of odor-active volatiles, *J. Food Sci.*, **52**, 375–377.

REISINGER, H. and MARR, R., 1992, Multicomponent-liquid-membrane permeation of organic acids, *Chem. Eng. Technol.*, **15**, 363–370.

REISINGER, H.H., MARR, R.J. and PREITSCHOPF, W., 1990, Liquid membranes in the field of biotechnology, *Proc. Int. Congr. Membranes and Membrane Processes*, Chicago, Vol. I, pp. 715–771.

RETHWISCH, D.G., SUBRAMANIAN, A., YI, G. and DORDICK, J.S., 1990, Enzyme facilitated transport and separation of organic acids through liquid membrane, *J. Am. Chem. Soc.*, **112**, 1649–1650.

SAMUEL, W.A. and LEE, Y.Y., 1980, *Biotech. Bioeng.*, **22**, 757–777.

SARTORI, L., MANOHAR, K., SIWAK, M. and SKEA, W., 1985, Isolation of cephalosporin-C from fermentation broths using membrane systems and HPLC, *Pharm. Manuf.*, 37.

SCHLICHER, L.R. and CHERYAN, M., 1990, Reverse osmosis of lactic acid fermentation broths, *J. Chem. Technol. Biotechnol.*, **49**, 129.

SCHLOSSER, S., 1993a, *J. Membr. Sci.*, **80**, 99–106.

SCHLOSSER, S., 1993b, Technical report of AJAC Project No. AIR3-CT94-1931.

SHEN, M.J., WILEY, R.C. and SCHLIMME, D.V., 1987, Solute and enzyme recoveries in apple juice clarification using ultrafiltration, *J. Food Sci.*, **52**, 732–736.

SHIMIZU, Y., SHIMODERA, K. and WATANABE, A., 1993, Cross-flow filtration of bacterial cells, *J. Ferment. Bioeng.*, **76**(6), 493–500.

STRATHMANN, H., 1986, *Electrodialysis, in Synthetic membranes: Science, Engineering and Applications*, D. Reidel, Dordrecht, pp. 197–223.

SZANIAWSKI, A.R. and SPENCER, H.G., 1997, Effects of immobilized pectinase on the microfiltration of dilute pectins solutions by microporous titania membranes: resistance model interpretation, *J. Membr. Sci.*, **127**, 69–76.

TANIGUCHI, M., KOTANI, N. and KOBAYASHI, T., 1987, High concentration cultivation of lactic acid bacteria in fermentor with cross-flow filtration, *J. Ferment. Technol.*, **65**(2), 179–184.

TIMMER, J.M.K. and KROMKAMP, J., 1994, Efficiency of lactic acid production by *Lactobacillus helveticus* in a membrane cell recycle reactors, *FEMS Microbiology Reviews*, **14**, 29–38.

TIMMER, J.M.K., VAN DER HORST, W.C. and ROBBERTSEN, T., 1993, Transport of lactic acid through reverse osmosis and nanofiltration membranes, *J. Membr. Sci.*, **85**, 205–216.

TULI, A., SETHI, R.P., KHANNA, P.K., MARWAHA, S.S. and KENNEDY, J.F., 1985, *Enzyme Microb. Technol.*, **7**, 164–168.

VAIJA, J., LINKO, Y.-Y. and LINKO, P., 1982, *Appl. Biochem. Biotechnol.*, **7**, 51–54.

VAN NISPEN, S.G.M. and JONKER, R., 1991, U.S. Patent 5,002,881.

VISACKY, V., STRATHMANN, H., DRIOLI, E., NAREBSKA, A., CUPERUS, F.P. and SCHLOSSER, S., 1996, Spacer for hybrid and multistage membrane process, Slovak patent application PV 0691–1996.

WAY, J.D., NOBLE, R.D., FLYNN, T.M. and SLOAN, E.D., 1982, Liquid membrane transport: a survey, *J. Membr. Sci.*, **12**, 239–259.

WIJIMANS, J.C., NAKAO, S. and SMOLDERS, C.A., 1984, Flux limitation in ultrafiltration: osmotic pressure model and gel layer model, *J. Membr. Sci.*, **20**, 115–124.

YASUTHOSHI, S., KENICHI, S. and ATSUO, W., 1993, Cross-flow microfiltration of bacterial cells, *J. Ferment. Biotechnol.*, **76**(6), 493–500.

4

Catalytic Membrane Reactors for Production of Pure Optically Active Compounds

4.1 Introduction

Chirality is the property which characterizes a pair of enantiomers (from Greek *enantio* = opposite), or molecules that are nonsuperimposable mirror images of each other (Figure 4.1). These molecules are diasymmetric and optically active (they can promote optical rotation). Therefore, enantiomers are also termed optical isomers and optical antipodes. The most common types of chiral molecules contain a tetrahedral carbon atom attached to four different groups; the carbon atom is the stereogenic (asymmetric) centre of the molecule.

Enantiomers have identical physical and chemical properties as long as they are in an achiral environment. Only when the system is chiral can the enantiomers be distinguished from each other.

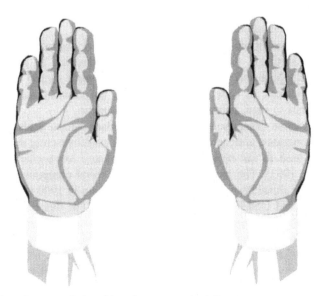

Figure 4.1 Mirror image relationship of asymmetrical figures.

Table 4.1 Different activities of enantiomers

Component	Effect of (S)-enantiomer	Effect of (R)-enantiomer
Dopa	Anti-Parkinson	Serious side effects
Penicillamine	Antiarthritic	Mutagen
Thalidomide	Teratogen	Sedative
Ketamine	Anaesthetic	Hallucinogen
Ethambutol	(S,S)- tuberculostatic	(R,R)- blindness
Novrad	(2R,3S)-(+) analgesic	(2S,3R)-(−) antitussive
Indacrinone	Uricosuric	Diuretic
Plactobutrazol	(2R,3S) plant growth regulation	(2R,3S) fungicide
Asparagine	Bitter	Sweet
Aspartame	(S,S) sweet	(R,R) bitter
Limonene	Lemon odour	Orange odour

Many biological processes, regulatory principles and biochemical reactions are based on chiral recognition phenomena. At the molecular level, asymmetry dominates biological processes. Chirality is not a prerequisite for bioactivity but in bioactive molecules where a stereogenic centre is present, great differences are usually observed for the activities of the enantiomers. This is a general phenomenon and applies to all bioactive substances, such as drugs, insecticides, herbicides, flavour compounds and food additives.

Enantiomers can have different biological activities, which often influence the efficacy or toxicity of the compound. Often one of the enantiomers is beneficially bioactive, whereas the other may be inactive, unwanted, or possibly toxic. For example, the racemate Verapamil has long been recognized as a calcium channel blocker for treatment of high blood pressure, angina and abnormal heart rhythms. The (S)-isomer alleviates high blood pressure even more effectively than the racemate, while the (R)-form is now found to inhibit resistance of cancer cells to anticancer drugs (Crosby, 1991). Other examples of different activities of enantiomers are listed in Table 4.1.

For these reasons, enantiomerically pure compounds are desirable and, in many cases, vital in the pharmaceutical, food and agrochemical industries. In fact, it has become evident that optically pure and well-defined chiral compounds should be used rather than mixtures of enantiomers, in order to better control the efficacy and consequences of administering chirals to man, animals and the environment.

Pure optically active compounds are used to produce amino acids, antibiotics, anti-inflammatories, hormones, anticancer drugs, vitamins, cardiovascular drugs, and so on. In Table 4.2, drugs used in the optically pure form (a) and as racemic mixtures (b) are reported (Sheldon, 1993). The growing need for pure chiral compounds is demonstrated by the growth of the chiral drug market from US$18 billion in 1990 to US$45.2 billion in 1994, and an estimated US$300 billion in the year 2000. For this reason, much effort is devoted to the study of technologies for production of pure enantiomers.

In the present chapter, catalytic membrane reactors and membrane processes applied to the production and separation of enantiomers will be described, following an introduction on the definition and properties of chirality and on the traditional alternative techniques used in the production of enantiomers. Membranes with intrinsic stereoselectivity have also been realized and will be discussed.

Table 4.2 Drugs used as the pure isomer and as racemic mixtures

Compound	Function
(a) Used as pure isomers	
Amoxicillin	Antibiotic
Ampicillin	Antibiotic
Captopril	ACE-inhibitor[a]
Enalapril	ACE-inhibitor[a]
Cefaclor	Antibiotic
Diltiazem	Calcium antagonist
Naproxen	NSAID
Cefalexin	Antibiotic
Lavastatin	Antihypercholesteremic
Cefratriaxone	Antibiotic
(b) Used as racemic mixtures	
Ibuprofen	NSAID
Atenolol	Beta-blocker
Albuterol	Bronchodilator
Iohexol	Contrast medium

[a] Angiotensin converting inhibitor.

4.2 Properties and Definitions of Chirality

The specific distinguishing physical property of enantiomers is the rotation of planc-polarized light. The degree (angle) of rotation is easily measured using a polarimeter and has a specific value for each optically active substance. If the rotation is to the right, the substance is *dextrorotatory* and is denoted by the symbol (+) or (*d*); if the rotation is to the left it is *levorotatory*, and it is denoted by the symbol (−) or (*l*). For optical activity to be exhibited, the energy barrier for conversion of a chiral molecule into its enantiomer must be high enough (> 80 kJ mol^{-1}) to allow isolation or at least observation of enantiomers.

The magnitude of the rotation of an enantiomer is reported as its specific rotation $[\alpha]$. This rotation is dependent on the wavelength of light used, the length of the polarimeter tube, the temperature, the solvent and the concentration. The monochromatic light most frequently used is the light emitted by the sodium lamp at 589 nm (the sodium D line). Thus, the specific optical rotation of a substance at 20 °C is given by

$$[\alpha]_D^{20} = \frac{\text{observed rotation (degrees)}}{\text{length of sample tube (dm)} \times \text{concentration (g ml}^{-1})}$$

The optical purity of a sample is the specific optical rotation of the particular enantiomeric mixture $[\alpha]$ divided by that of the pure enantiomer (the so-called absolute rotation) $[\alpha]_0$:

$$\text{optical purity} = \frac{[\alpha]}{[\alpha]_0}$$

However, this method is not considered the most useful to determine optical purity, since it compares samples with different chemical purities. For this reason, it is preferred to use the term enantiomeric excess (*ee*). For a case where $(R) > (S)$, the *ee* is given by

$$\text{enantiomeric excess } (ee) = \frac{(R - S)}{(R + S)} \times 100$$

Thus, a solution containing R and S at the ratio 90:10 has an *ee* of 80%.

The eudismic ratio (ER) (Lehmann et al., 1976) is defined as the ratio of the biological activity of the active enantiomer (eutomer) to that of the less active enantiomer (distomer: $ER = A_{eu}/A_{dis}$). The eudismic index (EI) is defined as $\log A_{eu} - \log A_{dis}$. ER and EI are measures of the stereospecificity of the substance. The higher is the ER the higher the efficiency of drug and the lower is the amount of drug needed.

The enantiomeric ratio (E) is a measure of the efficacy of a kinetic resolution. It is the ratio of the pseudo first-order rate constant for the two enantiomers; for example, for a reaction in which the (S)-enantiomer reacts selectively, the enantiomeric ratio is given by

$$E_S = \frac{k_S}{k_R}$$

The enantioproductivity (Q) is the ratio of the enantiomers formed, S/R. Q is directly proportional to the ratio of the rates of formation of the two enantiomeric products.

The separation factor of (R,S)-isomers is defined as follows:

$$\alpha = \frac{C_{pR}/C_{bR}}{C_{pS}/C_{bS}}$$

where C_{pR} and C_{bR} are the concentrations of isomer R in the permeate and bulk respectively; and C_{pS} and C_{bS} are the concentration of isomer S in the permeate and bulk.

The optical resolution ratio is expressed as the ratio between the flux of desired and undesired isomer. When (R) is the desired isomer, α is expressed as

$$\alpha_r = \frac{J_R}{J_S}$$

where J_R and J_S are the molar fluxes (mol m^{-2} h^{-1}) of (R) and (S) isomers, respectively.

With kinetic resolutions the enantiomeric ratio (E) can be used to compare quantitatively the efficiencies of different processes. To compare the efficiencies of different catalytic asymmetric syntheses, the enantioproductivity (Q) can be used. Thus, whereas in kinetic resolutions E is the ratio of the pseudo first-order rate constants of the two enantiomers, Q is directly proportional to the ratio of the rates of formation of the two enantiomeric products. Obviously, Q, in contrast to E, does not vary with conversion, since an asymmetric synthesis is a reaction of a single compound whereas a kinetic resolution is a reaction of a mixture. Q is a more useful parameter than *ee* for comparing the effect of different variables (e.g. catalyst, substrate, solvent) in asymmetric syntheses.

The following sections discuss traditional and innovative techniques.

4.3   Traditional Technology

Chiral chromatography, particularly ligand-exchange chromatography, has been developed as an effective analytical technique for separating enantiomers. High-performance liquid chromatography (HPLC) methods, which use chiral stationary phases, are used to separate enantiomers. Optically active ligands, such as amino acids, are covalently bound to a solid support, to form a chiral stationary phase. Amino acid derivatives such as *N*-(3,5-dinitrobenzoyl)phenylglycine are also used (Pirkle and Pochapsky, 1987).

Several serum proteins undergo enantioselective interactions with a wide variety of pharmacologically active compounds. This property has been used to develop chiral stationary phases that are based on bovine serum albumin bound to HPLC-grade silica. For a protein, the separation depends very much on parameters such as pH, ionic strength and temperature. For this reason, the optimum conditions have to be determined for each compound.

Polymers with a helical structure (such as cellulose derivatives and synthetic polymers) are able to separate enantiomers on the basis of steric effects. Cyclodextrins are cyclic oligosaccharides composed of α-D-glucose units linked through the 1,4-position. The three most common forms are α-, β-, and γ-cyclodextrin, containing respectively six, seven and eight glucose units (Figure 4.2). Cyclodextrins have a doughnut-shaped structure, the interior cavity of which is relatively hydrophobic. A variety of compounds fit into this cavity to form inclusion complexes. The β- and γ-forms have successfully been attached to silica to form chiral stationary phases that are widely used (Däppen et al., 1986; Wainer, 1984).

Chromatography is a process that results in high enantiomeric excess; it is suitable on the analytical scale but its scale-up can be difficult.

Production of enantiomers for optically pure drugs can be achieved by different routes. In many cases the pure compounds can be easily recovered by extraction techniques, from chiral (carbohydrates, terpenes, alkaloids) that occur naturally as pure enantiomers. Fermentation of inexpensive, abundantly available substrates, such as sucrose or molasses, is an important source of single chiral molecules, such as lactic, tartaric and L-amino acids, as well as for complex substances, such as antibiotics, hormones and vitamins (Buchta, 1983). Synthesis from either chiral or prochiral starting materials is also an important source of pure enantiomers.

The possible approaches for the preparation of optically pure compounds from inactive starting materials are asymmetric synthesis and resolution of racemates (Figure 4.3). Pure enantiomers can be obtained by asymmetric synthesis (Davies and Reider, 1996; Stinson, 1993; Davini, 1991), in which an enantiomeric reagent or catalyst is used to carry out a reaction on an achiral substrate (prochiral) to give a single chiral product.

In general, asymmetric synthesis is a selective technique, but it can become uneconomic if numerous steps and the use of expensive enantiomeric reagents are needed. In some cases, it is less expensive to produce a racemic mixture and then separate the enantiomers by physical methods, for example by diastereomeric crystallization or kinetic resolution. Diastereomeric crystallization is a technique whereby a covalent derivative is formed with an optically pure resolving agent; the different compounds obtained are separated and then decomposed to the corresponding enantiomers (Figure 4.4). This technique can be wasteful, because the unwanted isomer is often thrown away.

A kinetic resolution depends on the fact that the two enantiomers of a racemate react at different rates with a chiral catalyst, such as an enzyme. An asymmetric synthesis, on the other hand, involves the creation of an asymmetric (stereogenic) centre which, in this case, is by chiral discrimination of equivalent groups in an achiral starting material. Thus, a kinetic resolution involves substrate selectivity while an asymmetric synthesis involves product selectivity. Often, to discriminate between these two types of enantioselectivity, authors refer to the former as enantiospecific (specific for the substrate) and to the latter as enantioselective.

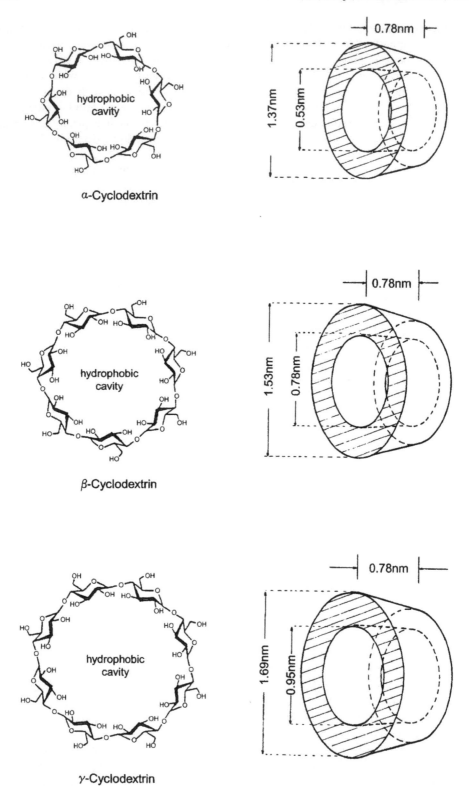

Figure 4.2 Structures of cyclodextrins.

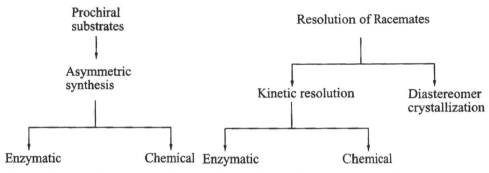

Figure 4.3 Possible routes for the preparation of optically pure compounds from inactive materials.

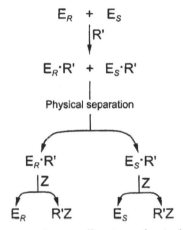

Figure 4.4 Scheme of diastereomeric crystallization: *classical optical resolution.* Racemic mixture of enantiomers (E_R; E_S) reacts with an enantiomer (R') and gives two diastereoisomers ($E_R \cdot R'$ and $E_S \cdot R'$). These can be separated by physical methods such as crystallization. Afterwards, the two diastereoisomers react with a component (Z) to regenerate each enantiomer (E_R; E_S) and the component R'Z.

Asymmetric synthesis has a theoretical yield of 100% compared to 50% for kinetic resolution. Nevertheless, kinetic resolutions have several advantages. They are simpler processes, often giving much better productivities for equivalent optical purities. Kinetic resolutions are generally employed for synthesizing the less reactive enantiomer of the substrate. The less reactive enantiomer can easily be obtained in high optical purity by carrying out the reaction to the appropriate conversion. In practice, it is often worthwhile sacrificing a few percent of chemical yield in order to obtain material of higher purity. This is not possible with asymmetric synthesis, since the reactions pertain to a simple (prochiral) substrate and enantioselectivity is (by definition) independent of conversion. The major disavantage of kinetic resolutions is that they require at least one extra step — racemization of the unwanted isomer. This can be circumvented if conditions can be found whereby the unwanted enantiomer undergoes spontaneous racemization *in situ*, leading to a kinetic resolution that is formally equivalent to an asymmetric synthesis (i.e. with a maximum yield of 100%) (Kimura et al., 1990).

The factors affecting the costs of processes for the production of pure enantiomers are the cost of substrate and resolving agent or (bio)catalyst; the chemicals used; the ease of racemization of the unwanted isomer; the total number of steps; and the position of the resolution step in the overall synthesis. The cost of substrate influences the choice of method (asymmetric synthesis or resolution of racemate), as well as the possibility of recycling the catalyst or the resolving agent. For economically viable processes, it is generally essential to obtain high chemical and optical yields at high concentration of substrate.

The case of racemization of the unwanted isomer is important because of the cost of the resolution process and because often it is the most difficult step in the overall process. Racemization is generally obtained treating the isomer with a strong acid or base at elevated temperatures. However, the forcing conditions required for acid- and base-promoted racemization often destroy the molecule. Racemase enzymes can be used to carry out racemization. This is an area that needs further systematic investigation.

The resolution (or asymmetric synthesis) step must be carried out as early as possible in the process. This is because every step performed on a racemate carries 50% ballast through the process. Removal of this ballast automatically halves the amount of reagents, solvents, reactor volume, and so on required in subsequent steps. Often, during a productive cycle, a resolution step via asymmetric catalysis (for example by enantioselective enzyme) of a racemic intermediate is introduced.

In addition to synthetic organic chemical approaches, asymmetric synthesis and resolutions of racemates, biotransformations have become key technologies. Nature, in fact, has long experienced the advantages of asymmetric synthesis and developed highly specific catalytic systems. Some companies have used biotransformation to produce chiral substances. For example, Celgene Corporation developed technologies for producing chiral amines using aminotransferase (Celgene Corporation, 1990) and β-hydroxy-α-amino acids using aldolase enzyme (Celgene Corporation, 1994).

It can be advantageous to combine biotransformation and synthetic organic technologies, as Lonza Corporation investigated for the production of (S)-amide. They used a nonstereoselective nitrile hydratase to catalyze conversion of the nitrile to racemic amide, and an amidase to hydrolyze the (R)-amide preferentially to the (R)-carboxylic acid, leaving the desired (S)-amide for Cilastatin production. The (R)-acid is then chemically converted to the racemic amide, which is combined with more racemic amide from the hydrolysis of nitrile, thus enabling a theoretical 100% conversion of starting material to the (S)-amide.

4.4 Membrane Chirotechnology

Membrane processes are being investigated for the development of new technologies to produce optically pure isomers and/or resolve racemic mixture of enantiomers.

On the basis of current literature there are basically two schemes for the use of membrane technology to produce enantiomers (Figure 4.5). In one case, the membrane itself is intrinsically enantioselective: the membrane is the chiral system that selectively separates the desired isomers on the basis of its spatial conformation. In the other, a kinetic resolution using an enantiospecific biocatalyst is combined with a membrane separation process; the membrane separates the product from the substrate on the basis of their relative chemical properties (i.e. solubility).

This kind of configuration is widely used to carry out kinetic resolutions of low water solubility substrates in biphasic membrane reactors (the detailed description of these reactors is reported in Chapter 6).

In the present section, catalytic membrane reactors and membrane processes applied for production and separation of enantiomers will be described.

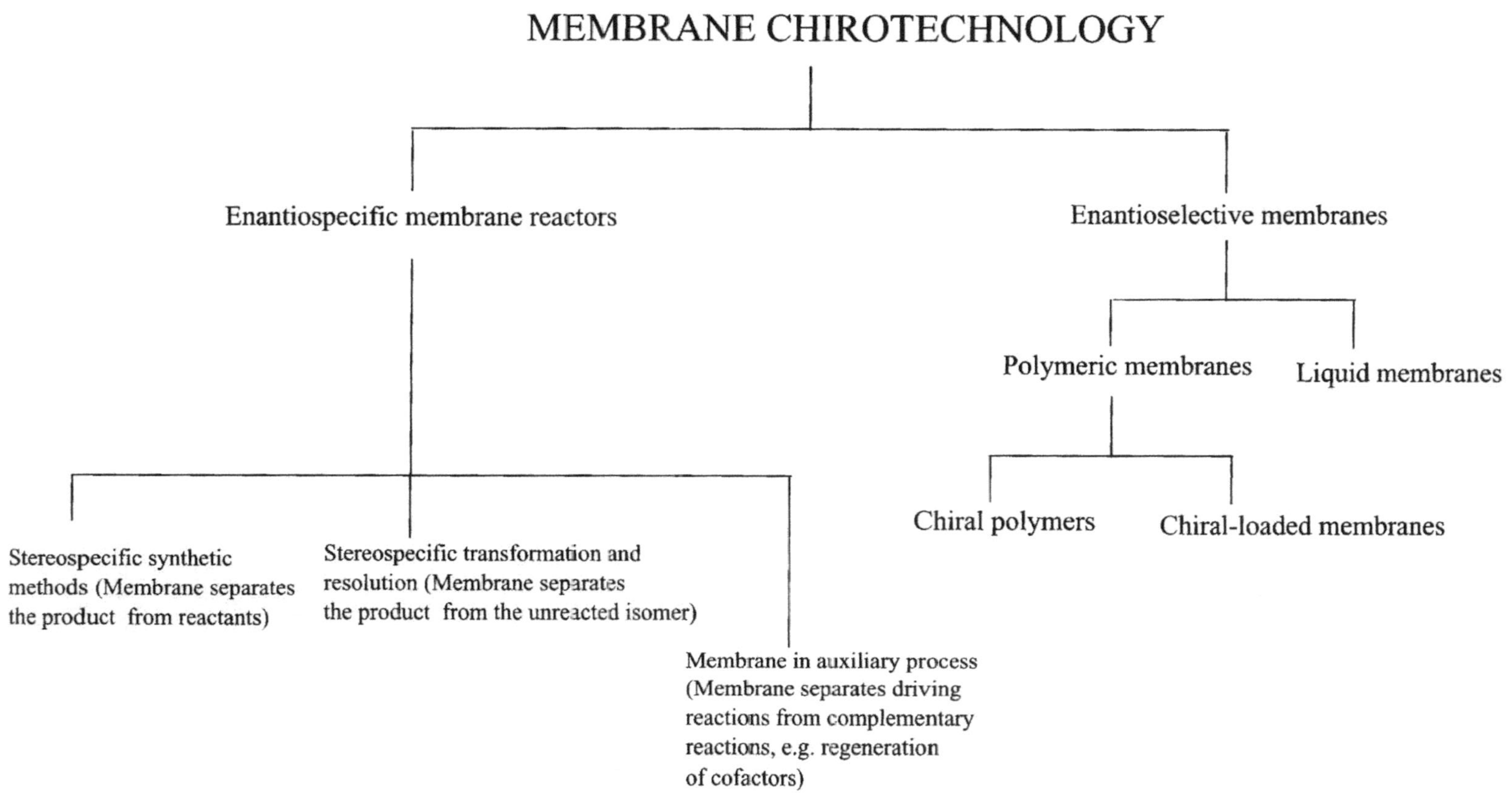

MEMBRANE CHIROTECHNOLOGY

Figure 4.5 Scheme of different strategies for membrane chirotechnology.

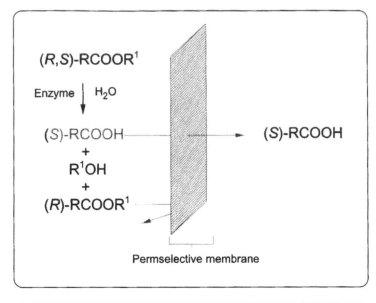

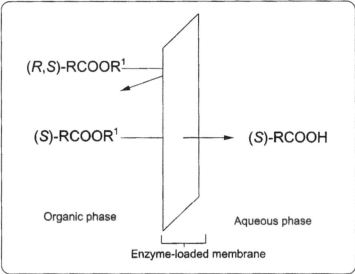

Figure 4.6 Membrane processes combined with enantioselective or enantiospecific systems.

When membrane separation processes are combined with enantioselective biotransformations (Figure 4.6) to realize simultaneously the bioconversion and the separation of the chiral compound, the chiral system is the enantiospecific enzyme and the membrane is a barrier that separates the preferentially converted isomer. The separation occurs thanks to the different solubility of product and substrate into the two phases.

4.4.1 *Enantiospecific Catalytic Membrane Reactors*

Competitive production of chiral compounds requires an inexpensive, large-scale process for producing and separating enantiomers. Membrane processes are suitable for this because they can work in continuous mode and large amounts of material can be processed.

Table 4.3 Catalysts for enantiospecific reactions

Catalyst	Application	Reference
L-Specific aminoacylase from *Aspergillum oryzal*	Production of of L-amino acids, such as L-methionine, L-valine	Chibata et al. (1987)
Amino acid esterase	Enantiospecific hydrolysis of amino acid esters	Meijer et al. (1985)
D-Specific hydantoinase from *Agrobacterium radiobacter*	Production of D-amino acid	Olivieri et al. (1981)
Racemases from		
Achromobacter obal	Racemization of D-α-amino-ε-caprolactans	Fukumura (1977)
Cryptococcus laurentii	Production of L-lysine	Fukumura (1977)
Amino acid dehydrogenases	Enantioselective reductive amination of α-keto acids to L-amino acids	See Chapter 5 (Table 5.5)
Pyridoxal-phosphate-dependent lyases and transferases	Synthesis of L-amino acids via carbon–carbon bond formation	Sheldon (1993)
Lipases and esterases	Stereospecific hydrolysis, esterification, etc.	See Chapter 6 (Figure 6.9)
Lipase	Production of (R)-α-phenoxypropionic acid	Kirchner et al. (1985); Fukui et al. (1990)
Lipase	Production of (S)-α-arylpropionic acid	McConville et al. (1990); Battistel et al. (1991)
Aminotransferase	Production of chiral amines	Celgene Corp. (1990)
Aldolase	β-hydroxy-α-amino acids	Celgene Corp. (1994)
Antibody	Hydrolysis of steroid esters	Kohen et al. (1980)
Antibody of 4-phenylphosphonate	Stereospecific reduction of α-ketoacids	Schultz and Nakayama (1992)
Monoclonal antibody 1C7	Stereoselective hydrolysis of esters	Nakatani et al. (1994)

Stereoselective hydrolases (lipases among others) are widely used and are especially attractive for large-scale preparations. This is because hydrolases are inexpensive and do not require cofactors (CMRs using cofactor-dependent enzymes are discussed in Chapter 5). Although in living systems lipases catalyze hydrolysis of triglycerides, they also have the capacity to interact with a variety of ester substrates, showing enantiospecificity for one isomer (Zhang and Wainer, 1993; Battistel et al., 1991; Roberts et al., 1992, 1993; Lopez and Matson, 1997).

Antibodies are also used to carry out stereoselective hydrolysis of esters (Nakatani et al., 1994). Catalytic antibodies attract much attention owing to their potential as novel protein catalysts with high selectivity, and to their similarities to enzymes in their catalytic behaviour. Since catalytic properties of antibodies were demonstrated in 1986, a variety of these applications has been explored (Kohen et al., 1980; Schultz and Nakayama, 1992; Nakatani et al., 1994).

A selection of the enantiospecific catalysts reported in the literature is summarized in Table 4.3.

Table 4.4 Enantiospecific membrane reactors

Reactions	Enantiospecific catalyst	Reactor type	Reference
Production of optically active 3-phenylglycidic acid ester	Lipase from *Serratia marcescens*	Ultrafiltration hollow-fibre membrane reactor	Matsumae et al. (1994)
Stereospecific production of (*S*)-(+)-ibuprofen	*Acinobacter* sp. AK 226	Ultrafiltration hollow-fibre membrane reactor	Takagi et al. (1994)
Production of L-phenylalanine	Dehydrogenases	Ultrafiltration membrane reactor	Schmidt et al. (1987)
(*S*)-Ibuprofen	Lipase from *C. cylindracea*	Immobilized enzyme membrane reactor	Zhang and Wainer (1993)
Production of D-phenylglycine-related amino acids	Hydantoinase	Polyacylamide gel	Yamada et al. (1980)
(2*R*,3*S*)-*trans* isomer of the methyl ester of 4-methoxy-3-phenylglycidic acid	Lipase entrapped in polyacrylonitrile capillary membranes	Biphasic membrane reactor	Lopez and Matson (1997)
Production of (*S*)-(+)-Naproxen	Lipase from *C. cylindracea* immobilized on amberlite	Packed-bed continuous bioreactor	Battistel et al. (1991)
Production of (*R*)-1-aminoidan	Subtilisin immobilized on glass beads	Continuous-flow column bioreactor	Gutman et al. (1992)

Ultrafiltration, membrane extraction, supported liquid membranes, and electrodialysis, are the most common membrane separation processes combined with biotransformations in realizing enantiospecific biocatalytic membrane reactors. A summary of enantiospecific membrane reactors is reported in Table 4.4.

Several studies have been carried out to optimize the production of amino acid enantiomers. Matson and Quinn (1979) studied the separation of L-amino acids from aqueous racemate solutions using an impregnated-liquid membrane in combination with an enzyme immobilized membrane.

The production of L-phenylalanine out of the racemic mixture of D,L-phenyllactate has been reported by Schmidt et al. (1987). The L-phenylalanine is obtained by two consecutive biotransformations catalyzed by D- and L-hydroxyisocaproate dehydrogenase and L-phenylalanine dehydrogenase. The reactions were carried out in an enzyme membrane reactor, where enzymes and cofactor (NAD/H) were compartmentalized behind an ultrafiltration membrane. The cofactor was covalently bound to polyethyleneglycol 20 000. The authors carried out kinetic modelling of the multienzyme-system and modelling of the membrane reactor operating as a continuous stirred tank reactor.

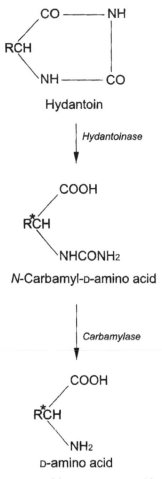

Figure 4.7 Schematic representation of bioconversion of hydantoin to D-amino acids.

The production of D-*p*-hydroxyphenylglycine (D-p-HPG) for example can be achieved by a subsequent enantioselective hydrolysis of 5-*p*-hydroxyphenylhydantoin (5-p-HPH) to *N*-carbamylhydroxyphenylglycine (N-carb-p-HPG) and to the corresponding amino acid as shown in Figure 4.7.

D-p-HPG is used for the production of semisynthetic penicillin and cefalosporin. Traditionally this D-amino acid was obtained by chemical synthesis. Recently, new processes have been developed, which consist of the asymmetric biotransformation of (D,L)-5-monosubstituted hydantoin followed by further biotransformation or chemical treatment of the derived *N*-carbamyl (Yamada et al., 1980).

Hydantoinases are present in many microorganisms, such as *Pseudomonas, Aerobacter, Agrobacter, Bacillus, Corinebacterium* and *Actinomycetes*. Hydantoinase from *Agrobacterium* was used to obtain the asymmetric biotransformation of D-p-HPH into the corresponding *N*-carbamyl (Giorno, 1988/89). The biocatalyst used was present in waste water from industrial production of optically pure *p*-HPG (provided by Recordati, Italy). Hydantoin is completely converted by hydantoinase and carbamylase (both present in *Agrobacterium*) into N-carb-p-HPG and D-p-HPG, respectively. In the industrial production of this D-amino acid, the carbamylase enzyme loses its activity more rapidly than

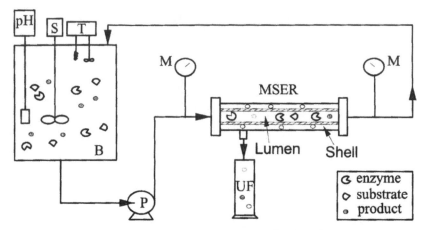

Figure 4.8 Membrane reactor system with enzyme confined in the lumen circuit. (B = batch; M = manometer; P = peristaltic pump; pH = pH meter; UF = ultrafiltered reservoir; S = stirrer; T = thermostat; MSER = membrane segregated enzyme reactor.)

hydantoinase. The recovery of hydantoinase activity present in the waste water and its reuse was of interest from the economic and environmental points of view.

The *N*-carbamyl obtained by biotransformation with hydantoin is optically pure and it can be converted to the corresponding D-amino acid by chemical hydrolysis with retention of configuration.

The enantioselective hydrolysis of 5-*p*-hydroxyphenylhydantoin into the corresponding *N*-carbamyl was studied in biocatalytic membrane reactors realized by segregating the *Agrobacterium* in the lumen circuit of a hollow-fibre reactor or by entrapping it within flat membranes by a phase-inversion method (Drioli et al., 1993).

In one case the microorganism containing the enzyme was segregated in the reaction environment by capillary membranes, as described in Figure 4.8. Ultrafiltration membrane modules (with nominal molecular mass cut-off of 10 kDa) were obtained from Romicon. Reactions were carried out at 40 °C, pH 8.50, 5-p-HPH 0.2 mol l^{-1}; 3.6 g l^{-1} of biomass, in 2 litres of total reaction volume. Under an applied transmembrane pressure of 7×10^4 Pa, solvent and reaction product permeated the membrane. In this way, the *N*-carbamyl was recovered in the permeate while the biocatalyst remained in the retentate and could be reused for further reactions by adding fresh substrate. The typical behaviour of biotransformation of hydantoin into *N*-carbamyl is illustrated in Figure 4.9, where the increase of product concentration as a function of time is shown.

The relative and absolute yields of product recovery for subsequent reactions and separation processes are reported in Table 4.5. The relative yield is defined as moles of product in the permeate/moles of total product, and absolute yield is defined as moles of product in the permeate/moles of substrate. The same sample of biomass could be reused for five subsequent experiments with a 20% decrease of activity per experiment. Each experiment lasted for 2 h. This means that 1 mole of 5-p-HPH could be converted in 10 h.

Catalytic membrane reactors have been studied with *Agrobacterium* entrapped in a flat-sheet membrane by phase inversion. Catalytic membranes were prepared using polysulfone (PS) as polymer, poly(vinyl pyrrolidone) (PVP) as water-soluble additive, *N*,*N*-dimethylformamide (DMF) as solvent, and lyophilized *Agrobacterium* cells as catalyst. The catalytic activity of biocatalyst-loaded membranes prepared with different percentages

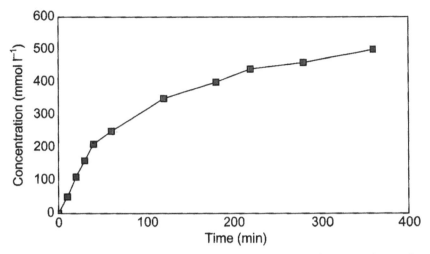

Figure 4.9 Typical behaviour of *N*-carbonyl concentration as a function of time during bioconversion of hydantoin.

Table 4.5 Relative and absolute yield of product recovery for consecutive reaction and separation processes in an enzymatic membrane reactor using hydantoinase

Run	Moles of product in permeate (*a*)	Moles of substrate (*b*)	Moles of product[a] (*c*)	Relative yield (*a/c*) (%)	Absolute yield (*a/b*) (%)
1	0.300	0.400	0.365	82	75
2	0.320	0.400	0.367	87	80
3	0.302	0.416	0.347	87	72
4	0.358	0.509	0.360	99	70
5	0.491	0.520	0.500	98	94

[a] Present in the permeate and the retentate.

of polymer and additive has been measured using small pieces of membrane suspended in a batch reaction system. In Table 4.6 the percentage of residual activity [(activity of immobilized biomass/activity of free biomass) × 100] is reported for each type of membrane prepared. Results are mostly affected by the percentage of PVP, which is soluble in water (used as non-solvent) and influences the porosity of the membrane during the phase-inversion process, and consequently the amount of immobilized cells. Organic solvents do not deactivate the microorganism. In fact, the specific activity of the microorganism was practically constant when it was kept in contact with the organic solvent for various times.

The initial reaction rate was higher when using biomass segregated in the hollow-fibre reactor than for biomass entrapped in the flat-sheet membrane. This is mainly due to low mass transport of substrate through the membrane to the catalyst that occurs when the catalyst is located inside the membrane wall.

The process described could be converted into a continuous process by recovering the *N*-carbamyl as it is produced. This could be achieved in a biphasic membrane reactor, where the hydantoin and the *Agrobacterium* are present in an organic solvent and the product is extracted in an aqueous phase.

Table 4.6 Percentage of residual activity for flat-sheet membranes loaded with *Agrobacterium*

% PS	% PVP	$IU/g_{\text{free biomass}}$	$IU/g_{\text{immob. biomass}}$[a]	Residual activity (%)
25	5	144	18.18	12.6
20	10	144	53.61	37
18	12	144	114.28	79
16	14	144	98.0	68

PS, polysulfone

PVP, poly(viny pyrrolidone)

[a] IU = international unit = μmol h^{-1}

Lopez and Matson (1997) reported the use of an enzyme membrane reactor for the production of a chiral intermediate used in the preparation of diltiazem, an important calcium channel blocker used in the treatment of hypertension and angina. This drug has two chiral centres and so it can potentially exist as any of four stereoisomers; since one of these isomers is considerably more efficacious than the others, the drug is marketed as the pure isomer.

The starting material used for diltiazem synthesis is a racemic mixture of the two (+)-*trans* enantiomers of the methyl ester of 4-methoxyphenylglycid acid (MMPG). To produce diltiazem with the correct stereoconfiguration, it is necessary to begin with the optically pure (2R,3S)-*trans* isomer of MMPG. This required (2R,3S)-*trans* enantiomer can be prepared by a biphasic subtractive resolution process wherein an enzyme in aqueous solution stereoselectively hydrolyzes the poorly water-soluble (2S,3S)-ester to the corresponding water-soluble acid, preferentially leaving most of the desired (2R,3S)-ester behind in the organic phase for recovery and subsequent conversion to diltiazem.

A multiphase/extractive membrane reactor facility was implemented that currently produces over 75 metric tonnes per year of optically pure (2R,3S)-glycidate ester (used commercially in the production of diltiazem).

Dodds and Lopez (1993) identified the enzyme able to stereoselectively hydrolyze the undesired (2S,3R)-glycidate methyl ester and thereby deplete it from the initially racemic ester pool. The enzyme used in the pilot-plant system was a lipase from a bacterial organism, *Serratia marcescens*. The principal advantage of this enzyme was the significantly higher enantioselectivity.

Multiphase membrane reactors offer some unique advantages in managing these sorts of reaction issues, as described in Chapter 6. The reactors examined contain hydrophilic and enzyme-activated porous membranes located at the interface between organic- and aqueous-phase process streams. The enzyme is entrapped within an asymmetric, solvent-resistant hollow-fibre membrane. The membrane is housed in stainless steel hollow-fibre membrane modules fabricated with a solvent-resistant epoxy potting compound. Bench-scale modules contained a nominal active membrane area of 0.75 m^2. The larger modules designed for pilot-plant and production use contained 7.5 and 60 m^2 of active membrane area, respectively. Unlike their laboratory-scale counterparts, these larger units were based on a cartridge-in-a-shell design than facilitated membrane replacement at required intervals.

This asymmetric membrane structure effectively 'immobilizes' the enzyme by localizing it between two barriers: the enzyme-impermeable interior skin, and the aqueous/

organic interface maintained at the exterior surface of the hollow fibre. The size of the enzyme prevents it from diffusing across the skin, while the low organic-phase solubility of the enzyme prevents it from partitioning into and being carried away by the organic process stream.

Few enzymes retain full activity for very long under the relatively harsh conditions characteristic of industrial use. Reversible enzyme containment in a hollow-fibre membrane simplifies replacement of deactivated enzyme in the field. Once enzyme activity has diminished excessively, old enzyme can be replaced by back-flushing and then reloading.

The development process was followed from bench-scale studies of process feasibility through optimization, process reliability, and pilot-plant studies, a process that ultimately culminated in the operation of a commercial-scale membrane reactor facility that currently produces over 75 metric tonnes per year of diltiazem intermediate.

A commercial-scale multiphase/extractive membrane reactor system for the production of optically pure (2R,3S)-methylmethoxyphenylglycidate ester was installed in Japan. The full-scale facility comprises two banks of twelve 60 m^2 production-scale modules for a total effective membrane reactor area of 1440 m^2. The annualized plant production rate is 53 kg per m^2 per year of 99% *ee* ester which, given the 5 day per week operating schedule, is equivalent to a membrane reactor productivity of more than 75 kg per m^2 per year based on actual membrane reactor run time.

This commercial membrane reactor installation has now been operating for four years and currently produces over 75 metric tonnes per year of the resolved 'GLOP' intermediate, from which some 110 tonnes per year of diltiazem is produced for sale in the United States. At present, material produced from membrane reactors accounts for more than half of the diltiazem now being sold in the United States as Cardizem.

4.4.2 *Intrinsically Enantioselective Membranes*

Optical substances can only be separated on the basis of physical stereoselectivity. In order to show optical resolution, membranes must contain a chiral microenvironment in their structure. Polymeric membranes with intrinsic enantioselective properties can be prepared using chiral polymers or by chiral modification of achiral porous membranes using chiral recognition agents, such as amino acid enantiomers, cyclodextrins (Figure 4.2), chiral crown ethers (Figure 4.10), oligopeptides, antibodies, cyclophanes or calixarenes. Cyclodextrins and crown ethers are useful selectors for this application. This is probably because the chiral rings of these molecules are able to host different chiral molecules. Owing to the hydrophilic external surface of the chiral ring of cyclodextrins they are water soluble, and because of their hydrophobic inner surface they may complex and carry over polar molecules.

The organic-soluble crown ethers, having nonpolar character for the outer surface of the chiral ring and polar for the inner one, are able to transport polar molecules from organic phase to aqueous phase.

These chiral agents can also be used to realize liquid membranes or supported liquid membranes. Enantiospecific polymers, prepared mainly for chromatographic purposes, are potentially applicable for direct membrane enantioseparation. Thus, optically active polyacryl- and polymethacryl amides, and cellulose derivatives could be used to prepare enantioselective membranes. Polymers modified with cyclodextrins have been used to prepare chiral membranes by phase-inversion methods; alternatively, the cyclodextrin

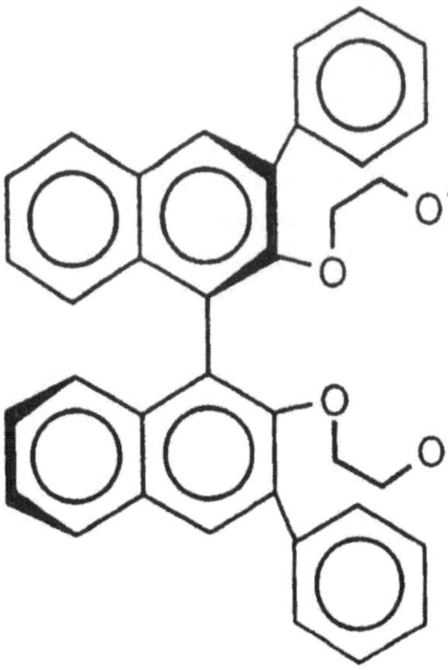

Figure 4.10 A chiral crown ether.

has been entrapped within the membrane during the phase-inversion process. A list of intrinsically enantioselective membranes is given in Table 4.7. Inoue et al. (1997) reported the preparation of hexa-armed poly(L-glutamates) with di- and tri(ethylene glycol) monomethyl ether units and their functionalities as enantioselective membranes for tryptophan, phenylalanine and tyrosine.

The enantioselectivity of chiral membranes can be affected strongly not only by the chemical structure but also by other factors, such as morphology and microcrystalline structure of polymers.

Polymers having enantiospecific cavities can be prepared by the copolymerization of an optically active template molecule of very similar structure to the molecule to be resolved. After the membrane is formed, the template molecule can be split off, leaving imprinted in the macroporous structure the structure of the molecule to be separated (Figure 4.11).

Masawaki et al. (1992) prepared enantioselective membranes from a solution of polysulfone (PS, as polymer), N-methyl-2-pyrrolidone (NMP) as solvent, $LiNO_3$ as swelling agent, and L-phenylalanine condensate with glutaraldehyde as chiral recognition agent. The selective transport of the L-amino acid through the L-phenylalanine condensate membrane is based on the more specific and stable interactions between L-configurations than between L- and D-forms. In this kind of membrane the polymeric matrix is made of polysulfone. This part of the membrane passes both D- and L-phenylalanine. The general result of the separation process will be preferential permeation of the L-phenylalanine with respect to the D-phenylalanine.

The membrane sheet was placed between two compartments of a dialysis cell. The aqueous (D,L)-phenylalanine was resolved with a separation factor of $\alpha_0 = 1.25$ using a concentration of 1 mmol cm^{-3} and pressure gradient of 0.05 MPa across the membrane. At higher operating pressure, the increasing volume flux decreased the enantioselectivity.

Polymeric membranes for optical resolution have been prepared using molecular imprinting by Yoshikawa et al. (1996) using butoxycarbonyl-L-tryptophan (Boc-L-Trp) as imprinting molecule, tetrahydrofuran (THF) as solvent, a polystyrene resin bearing a

Table 4.7 Enantioselective membranes

Membrane operation	Chiral selector	Compound separated	Reference
Membrane-based solvent extraction (MBSX) through hollow fibre	Copper(II) *N*-dodecyl-(L)-hydroxyproline	Leucine	Ding et al. (1992)
Supported liquid membrane	Chiral crown ethers	Phenylglycine, methionine, phenylalanine	Yamaguchi et al. (1985a,b) Shinbo et al. (1992)
Supported liquid membrane	Nopol; (2*S*)-(−)-2-methyl-1-butanol	Amino acids	Bryjak et al. (1993)
Supported liquid membrane and MBSX	(*S*)-*N*-(1-naphthyl)leucine octadecyl ester in dodecane	α-Amino acid esters α-Amides	Pirkle and Pochapsky (1987) Pirkle and Bowen (1994)
Micellar enhanced ultrafiltration	(L)-5-Cholesterol glutamate	Amino acids	Creagh et al. (1994)
Ultrafiltration	Molecularly imprinted polymeric membranes (DIDE derivatives)	Tryptophan, phenylalanine, alanine	Yoshikawa et al. (1996)
Ultrafiltration	L-Phenylalanine condensate with glutaraldehyde in polymeric membranes	Phenylalanine	Masawaki et al. (1992)
Ultrafiltration	*l*-Menthol polymerized onto polymeric membrane	Tryptophan, phenylalanine, tyrosine	Tone et al. (1995)
Ultrafiltration	Poly{γ-[3-(pentamethyldisiloxanyl)-propyl]-L-glutamate} membrane	Tryptophan	Aoki et al. (1995)
Ultrafiltration	Membranes prepared from 3α-helix bundle polyglutamates with oxyethylene chains	Tryptophan, phenylalanine, tyrosine	Inoue et al. (1997)
Ultrafiltration	BSA condensed with glutaraldehyde in polymeric membranes	Amino acids, such as phenylalanine	Higuchi et al. (1994)

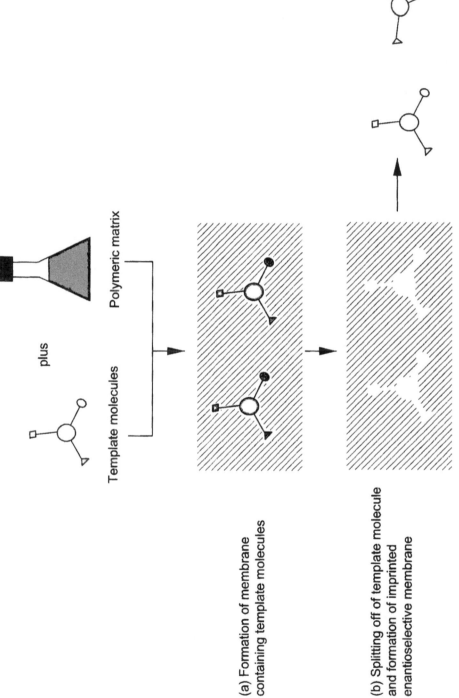

plus

Template molecules   Polymeric matrix

(a) Formation of membrane
containing template molecules

(b) Splitting off of template molecule
and formation of imprinted
enantioselective membrane

Figure 4.11 Preparation of enantioselective molecularly imprinted polymeric membrane.

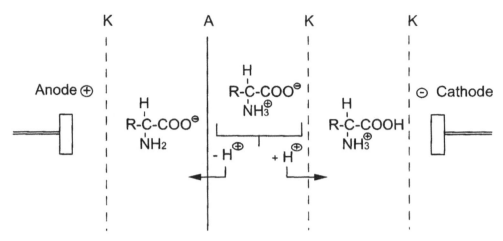

Figure 4.12 Separation of amino acid enantiomers by electrodialysis.

tetrapeptide derivative, H-Asp-(OcHex)-Ile-Asp(OcHex)-Glu(OBzl)-CH2-(DIDE-resin), as recognition site, and acrylonitrile and styrene as copolymer. After the membrane was formed, the imprinting molecule was removed using methanol. The molecularly imprinted membrane showed selective adsorbtion of the L-Trp (imprinting molecule) over the D-isomer. Electrodialysis was used to selectively permeate the isomer, which is preferentially adsorbed in the membrane.

Ultrafiltration membranes were prepared also from poly(γ-[3-(pentamethyldisiloxanyl)-propyl]-L-glutamate) (Aoki et al., 1995). This polymer has an α-helical structure in the membrane. Optical resolution of ($\pm$)-tryptophan was obtained by pressure-driven permeation through the membrane. The enantioselectivity was 16% enantiomeric excess for more than 160 h.

Electrodialysis is another suitable technique for the separation of amino acids. The principle of electrodialysis of an amino acid enantiomer is illustrated in Figure 4.12. Amino acids are amphoteric electrolytes, with both anionic and cationic groups in the molecule. The permeability of an amino acid through an ion exchange membrane is determined by the degree of dissociation of the carboxylic or amino group, which is influenced by the pH. Electrodialysis can be used as a chiral resolution system if the membrane has a chiral recognition site, or as a process combined with asymmetric biosynthesis. Both systems have been investigated (Yoshikawa et al., 1997).

The possibility of modifying commercially available achiral porous membranes with optically active selectors is also attractive. In this case, the chiral selector is 'loaded' onto the membrane by chemical or physical methods.

The methods described for immobilization of enzymes can be used when loading antibodies, peptides, amino acids, etc., as chiral selectors. The chiral molecules promoting the enantioselective mass transport are optically pure molecules able to undergo very well-defined enantiospecific interactions. There is no mobility of the complexing molecules; they are anchored on the achiral membrane surface and pores. The enantioselective mass transport is obtained by subsequent interactions (complexation/dissociation) between the molecule to be separated and the chiral agent present throughout the membrane. The distribution and concentration of chirals in the porous membrane are parameters that govern the selectivity and permeability of the membrane. Higuchi et al. (1994) immobilized bovine serum albumin (BSA) by means of glutaraldehyde onto polysulfone ultrafiltration membranes to resolve racemic phenylalanine. Depending on the buffer solution used, a high content of D-isomer was obtained in the permeate.

Supported liquid membranes (SLM) containing chiral recognition sites (chiral carrier) have also been used for separation of amino acid enantiomers. In these membranes the chiral system is constituted by the carrier, which transports the selected enantiomer from a source phase to a receiving phase. The organic solvent in which the carrier is diluted does not have optical resolution capacity, and both D- and L-isomer can diffuse through it. In order to have a good optical resolution ratio (α) through a SLM, it is advisable to use a highly stereoselective carrier diluted in an organic solvent through which non-carrier-mediated diffusion of D,L-isomers does not occur or at least is very low. In fact, if we consider a SLM with a carrier specific for the L-isomer, the optical resolution ratio is given by

$$\alpha = J_L/J_D$$

where: J_L is the flux of L-isomer mediated by the carrier plus the flux of L-isomer diffused through the solvent, and J_D is the flux of D-isomer mediated by the carrier plus the flux of D-isomer diffused through the solvent. Since diffusion of D- and L-isomers through the solvent is the same, a low optical resolution ratio is obtained, independently of the use of a highly selective carrier, if this transport is not negligible with respect to that mediated by the optical carrier.

In general, the parameters that influence the mass transport through a SLM are primarily the viscosity of organic solvent; the viscosity, molecular size and concentration of the organic carrier and solute to be transported; the temperature; and the solubility of the solute in the different phases.

Although SLMs have numerous advantages as a separation technique, they have not found application at industrial level. This is mostly because the stability of SLMs has not been sufficiently improved. Various studies on SLM stability are available in the literature (Danesi et al., 1987; Takeuchi et al., 1987; Chiarizia, 1991).

Enantiomers of D,L-phenylglycine have been resolved by liquid membranes containing chiral crown ethers (Yamaguchi et al., 1985a, 1985b, 1988). The authors studied the transport rate and enantioselectivity of several α-amino acids and amines through these kind of membranes (Shinbo et al., 1992). The highest separation was obtained for phenylglycine, with an optical resolution factor (α) of 18.9; followed by glutamic acid ($\alpha = 15.0$); arginine ($\alpha = 14.7$); methionine ($\alpha = 13.6$); and leucine ($\alpha = 13$).

Membrane-based solvent extraction has also been used to separate amino acids from a feed to an enriched phase. Pirkle and Bowen (1994) used chiral selectors derived from N-(1-naphthyl)leucine to separate amino acid enantiomers. The mode of operation of the system is shown in Figure 4.13. The stripping phase containing the chiral selector is confined in two membrane modules in contact with a source phase on one side, and with a receiving phase on the other.

An aqueous two-phase extraction system has been applied to the resolution of tryptophan enantiomers using bovine serum albumin as the chiral selector (Ekberg and Sellergen, 1985).

Industrial applications of these technologies are limited by the long processing times required because of the small contact area between the ligand-containing phase and the aqueous amino acid mixture.

Racemic mixtures of (D,L)-phenylalanine have been resolved by micellar-enhanced ultrafiltration (Creagh et al., 1994). The principle of separation is shown in Figure 4.14. A mixture of D,L-phenylalanine is fed into an ultrafiltration cell (UF membrane of 5 kDa molecular mass cut-off) that contains micelles containing chiral ligand. Because the

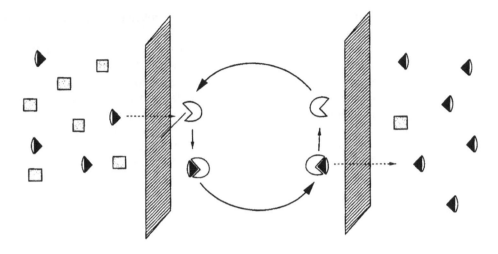

Legend

◖ Wanted enantiomer;

▢ Unwanted enantiomer;

◖ Chiral carrier

Figure 4.13 Operation of enantioseparation system by membrane-based solvent extraction.

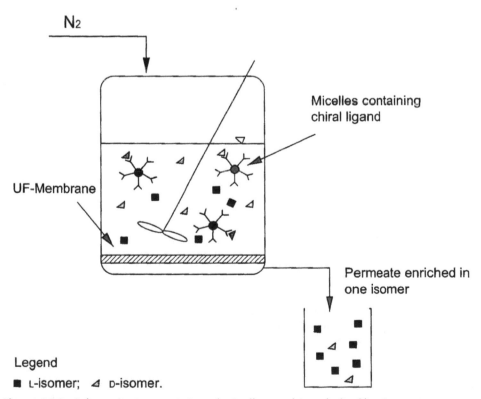

Legend

■ L-isomer; ◁ D-isomer.

Figure 4.14 Schematic representation of micellar emulsioned ultrafiltration system.

D-Phe preferentially adsorbs to the micelles, the permeate collected from the UF cell is enriched in L-phenylalanine.

Chiral emulsion liquid membranes (ELM) have also been used to resolve enantiomers. The advantages of emulsion liquid membrane extraction include fast transport and high capacity for polar solutes, common to most liquid membrane processes. In addition, ELM are also inherently less susceptible to disruption by surface-active agents, such as chiral carriers, owing to the stabilizing effect of the surfactants used in their formation. For this reason, they are more stable than supported liquid membranes (Zha et al., 1995).

Pickering and Chaudhury (1997) have studied resolution of (D,L)-phenylalanine by ELM using copper(II) N-decyl-L-hydroxyproline as a chiral carrier in a hexanol–decane membrane solvent. They observed a maximum enantioselectivity of 2.4.

4.5 Conclusions

The different methodologies that can be considered by a researcher when embarking on the industrial synthesis of a pure enantiomer have been discussed.

It is not possible to draw general conclusions about the superiority of one type of technology compared to another. The most economic technique depends on the components. For this reason, each case must be investigated individually.

In general, the chiral step should be introduced as early as possible in the overall process, but this consideration may be countered by other factors, such as racemization of the unwanted isomer.

Membrane chirotechnology is an emerging technique that can offer several advantages in terms of productivity, purity of simple isomers, and ease of scale-up. The number of papers published indicates the effort put into the use of membrane techniques. Until now, these techniques have been used mainly at laboratory scale. Real application at larger scale needs more investment, particularly in terms of development of experimental set-up rather than investigations on chirality, which has been investigated in the chromatographic field.

4.6 References

AOKI, T., TOMIZAWA, S. and OIKAWA, E., 1995, Enantioselective permeation through poly-γ-[3-(pentamethyldisiloxanyl)propyl]-L-glutamate membranes, *J. Membr. Sci.*, **99**, 117–125.

BATTISTEL, E., BIANCHI, D., CESTI, P. and PINA, C., 1991, Enzymatic resolution of (S)-(+)-naproxen in a continuous reactor, *Biotechnol. Bioeng.*, **38**, 659–664.

BRYJAK, M., KOZLOWSKI, J., WIECZOREK, P. and KAFARSKI, P., 1993, Enantioselective transport of amino acid through supported liquid membranes, *J. Membr. Sci.*, **85**, 221–227.

BUCHTA, K., 1983, *Biotechnology*, 3, pp. 410–420, Verlag Chemi, Weinheim.

CELGENE CORPORATION, 1990, U.S. Patent No. 4,950,606.

CELGENE CORPORATION, 1994, U.S. Patent No. 5,346,828.

CHIARIZIA, R., 1991, Stability of supported liquid membranes containing long-chain aliphatic amines as carriers, *J. Membr. Sci.*, **55**, 65.

CHIBATA, I., TOSA, T. and SATO, T., 1987, in REHM, H.J. and REED, G. (eds) *Biotechnology, 7*, pp. 653–684, VCH, Weinheim.

CREAGH, A.L., HASENACK, B.B.E., VAN DER PADT, A., SUDHLTER, E.J.R. and VAN'T RIET, K., 1994, Separation of amino-acid enantiomers using micellar-enhanced ultrafiltration, *Biotechnol. Bioeng.*, **44**, 690–698.

CROSBY, J., 1991, Synthesis of optically active compounds: a large scale perspective, *Tetrahedron*, **47**, 4789–4846.

DANESI, P.R., REICHLEY-YINGER and RICKERT, P.G., 1987, Lifetime of supported liquid membranes: the influence of interfacial properties, chemical composition and water transport on the long-term stability of the membranes, *J. Membr. Sci.*, **31**, 117.

DÄPPEN, R., ARM, H. and MEYER, V.R., 1986, *J. Chromatogr.*, **373**, 1–20.

DAVIES, I.W. and REIDER, P., 1996, Practical asymmetric synthesis, *Chemistry & Industry*, 412–413.

DAVINI, E., 1991, Synthesis of bioactive compound intermediates from chiral natural precursors, *Chemistry Today*, 41–47.

DING, H.B., CARR, P.W. and CUSSLER, E.L., 1992, Racemic leucine separation by hollow fibre extraction, *AIChE J.*, **38**, 1493.

DODDS, D.R. and LOPEZ, J.L., 1993, Enzymatic hydrolysis of glycidate esters in the presence of bisulfite anion, US patent 5,274,300.

DRIOLI, E., GIORNO, L., DONATO, L., MOLINARI, R. and BASILE, A., 1993, Membrane operation in biochemical processing, in *Chemistry and properties of biomolecular systems*, **2**, 193–204, N. Russo, J. Anastassopuolou and G. Barone Eds. Kluwer, Dordrecht.

EKBERG, B. and SELLERGREN, B., 1985, Direct chiral resolution in an aqueous two-phase system using the countercurrent distribution principle, *J. Chromatogr.*, **333**, 211–214.

FUKUI, T., KAWAMOTO, T., SONOMOTO, K. and TANAKA, A., 1990, Long-term continuous production of optically active 2-(4-chlorophenoxy)propanoic acid by yeast lipase in an organic solvent system, *Appl. Microbiol. Biotechnol.*, **34**, 330–334.

FUKUMURA, T., 1977, *Agr. Biol. Chem.*, **41**, 1327.

GIORNO, L., 1988/89, *Analisi cinetica di biocatalizzatoni immobilizzati o compartimentalizzati in membrane polimeriche*, Master Thesis, University of Calabria, pp. 57–113.

GUTMAN, A.L., MEYER, E., KALERIN, E., POLYAK, F. and STERLING, J., 1992, *J. Chromatogr.*, **40**, 760–767.

HIGUCHI, A., HARA, M., HORINGHI, T. and NAGAKAWA, T., 1994, *J. Membr. Sci.*, **93**, 157.

INOUE, K., MIYAHARA, A. and ITAYA, T., 1997, Enantioselective permeation of aminoacids across membranes prepared from 3α-helix bundle polyglutamates with oxyethylene chains, *J. Am. Chem. Soc.*, **119**(26), 6191–6192.

KIMURA, M., OHKUMA, T., TOKUNAGA, M. and NOYORI, R., 1990, *Tetrahedron Asymmetry*, **1**, 1–4.

KIRCHNER, G., SCOLLAR, M.P. and KLIBANOV, A.M., 1985, *J. Am. Chem. Soc.*, **107**, 7072–7076.

KOHEN, F., KIM, J.B., BARNARD, G. and LINDER, H.R., 1980, Antibody-enhanced hydrolysis of steroid esters, *Biochem. Biophys. Acta*, **62**(9), 328–337.

LEHMANN, P.A., RODRIGUES DE MIRANDA, J.F. and ARIENS, E.J., 1976, *Prog. Drug Res.*, **20**, 101–142.

LOPEZ, J.L. and MATSON, S.L., 1997, A multiphase/extractive enzyme membrane reactor for production of diltiazem chiral intermediate, *J. Membr. Sci.*, **125**, 189–211.

MASAWAKI, T., SASAI, M. and TONE, S., 1992, Optical resolution of an amino acid by an enantioselective ultrafiltration membrane, *J. Chem. Eng. Jpn.*, **25**(1), 33–39.

MATSON, S.L. and QUINN, J.A., 1979, *AICHE Annual Meeting*, San Francisco.

MATSUMAE, H., FURUI, M., SHIBATANI, T. and TOSA, T., 1994, Production of optically active 3-phenylglycidic acid ester by the lipase from *Serratia marcescens* on a hollow-fibre membrane reactor, *J. Ferment. Bioeng.*, **78**(1), 59–63.

MCCONVILLE, F.X., LOPEZ, J.L. and WALD, S.A., 1990, in AMBRAMOWICZ, D.A. (ed.), *Biocatalysis*, pp. 167–177, Van Nostrand Reinold, New York.

MEIJER, E.M., BOESTEN, W.H.J., SCHOEMAKER, H.E. and VAN BALKEN, J.A.M., 1985, in TRAMPER, E.J., VAN DER PLAS, H.C. and LINKO, P. (eds) *Biocatalysts in Organic Synthesis*, pp. 135–156, Elsevier, Amsterdam.

NAKATANI, T., UMESHITA, R., HIRATAKE, J., SHINZAKI, A., SUZUCHI, T., NAKAJIMA, H. and ODA, J., 1994, Characterization of a catalytic residue and mode of product inhibition, *Bioorg. Med. Chem.*, **2**(6), 457–468.

OLIVIERI, R., FASCETTI, E., ANGELINI, L. and DEGEN, L., 1981, *Biotechnol. Bioeng.*, **23**, 2173.

PICKERING, P.J. and CHAUDHURI, J.B., 1997, Enantioselective extraction of (D)-phenylalanine from racemic (D,L)-phenylalanine using chiral emulsion liquid membranes, *J. Membr. Sci.*, **127**, 115–130.

PIRKLE, H.W. and BOWEN, W.E., 1994, Preparative separation of enantiomers using hollow-fibre membrane technology, *Tetrahedron: Asymmetry*, **5**(5), 773–776.

PIRKLE, W.H. and POCHAPSKY, T.C., 1987, *Adv. Chromatogr.*, **27**, 73–78.

ROBERTS, S.M., WIGGINS, K. and CASY, G., 1992, *Preparative Biotransformations: Whole Cell and Isolated Enzymes in Organic Synthesis*, Wiley, New York.

ROBERTS, S.M., TURNER, N.J. and WILLETS, A.J., 1993, Some recent advances in the synthesis of optically pure fine chemicals using enzyme-catalyzed reactions in the key step, *Chemistry Today*, 9–17.

SCHMIDT, E., VASIC-RACKI, D. and WANDREY, C., 1987, enzymatic production of L-phenylalanine from the racemic mixture of D,L-phenyllactate, *Appl. Microbiol. Biotechnol.*, **26**, 42–48.

SCHULTZ, P.G. and NAKAYAMA, G.R., 1992, *J. Am. Chem. Soc.*, **114**, 780.

SHELDON, R.A., 1993, Chirality and biological activity, in *Chirotechnology*, pp. 39–72, Marcel Dekker, New York.

SHINBO, T., YAMAGUCHI, T., SAKAKI, K., YANAGISHITA, H., KITAMOTO, D. and SUGIURA, M., 1992, Enantioselective transport of amino acids and amines through a supported liquid membrane containing chiral crown ether, *Chemistry Express*, 7(10), 781–784.

STINSON, S.C., 1993, Chiral drugs, *Chem. Eng. News*, 38–65.

TAKAGI, M., OISHI, K., ISHIMURA, F. and FUJIMATSU, I., 1994, Production of S-(+)-ibuprofen from nitrile compound by enzymatic reaction combined with ultrafiltration, *J. Ferment. Bioeng.*, 54–58.

TAKEUCHI, H., TAKAHASHI, K. and GOTO, W., 1987, Some observations on the stability of supported liquid membranes, *J. Membr. Sci.*, **33**, 19.

TONE, S., HASAWAKI, T. and HAMADA, T., 1995, The optical resolution of amino acids by ultra-filtration membranes fixed with plasma polymerized *l*-menthol, *J. Membr. Sci.*, **103**, 57–63.

WAINER, I.W. and DOYLE, T.D., 1984, *J. Chromatogr.*, **284**, 117–122.

YAMADA, H., SHIMIZU, S., SHIMADA, H., TANI, Y., TAKAHASHI, S. and OHASHI, T., 1980, Production of D-phenylglycine-related amino acids by immobilized microbial cells, *Biochimie*, **62**, 395–399.

YAMAGUCHI, T., NISHIMURA, K., SHINBO, T. and SUGIURA, M., 1985a, Enantiomer resolution of amino acids by a polymer-supported liquid membrane containing a chiral crown ether, *Chem. Lett.*, 1549–1552.

YAMAGUCHI, T., NISHIMURA, K., SHINBO, T. and SUGIURA, M., 1985b, Chiral crown ether-mediated transport of phenylglycine through an immobilised liquid membrane, *Maku (Membrane)*, **10**, 178.

YAMAGUCHI, T., NISHIMURA, K., SHINBO, T. and SUGIURA, M., 1988, Amino acid transport through supported liquid membranes: mechanism and its application to enantiomeric resolution, *Bioelectrochem. Bioenerg.*, **20**, 109–123.

YOSHIKAWA, M., IZUMI, J., KITAO, T. and SAKAMOTO, S., 1996, Molecularly imprinted polymeric membranes containing DIDE derivatives for optical resolution of amino acids, *Macromolecules*, **29**, 8197–8203.

YOSHIKAWA, M., IZUMI, J. and KITAO, T., 1997, Enantioselective electrodialysis of amino acids with charged polar side chains through molecularly imprinted polymeric membranes containing DIDE derivatives, *Polymer J.*, **29**(3), 205–210.

ZHA, F.F., FANE, A.G. and FELL, C.J.D., 1995, Effect of surface tension gradients on stability of supported liquid membranes, *J. Membr. Sci.*, **107**, 75.

ZHANG, X-M. and WAINER, I.W., 1993, On-line determination of lipase activity and enantio-selectivity using an immobilized enzyme reactor coupled to a chiral stationary phase, *Tetrahedron Lett.*, **34**(30), 4731–4734.

5

Catalytic Membrane Reactors for Retention and Recycling of Coenzyme

5.1 Introduction

Important enantiospecific biotransformations are catalyzed by coenzyme-dependent oxidoreductases. The industrial application of these enzymes is limited by the high cost and instability of the coenzyme. They are relatively low molecular mass molecules and are difficult to recover and reuse after reaction has been completed. On the other hand, the high added value fine chemicals that can be obtained with oxidoreductase enzymes encourages investigations in this field.

Studies are oriented to the development of new technologies able to retain and recycle coenzyme in the reaction mixture, which will make reliable the use of these biocatalytic systems on large scale.

In the present chapter, catalytic membrane reactors for carrying out biotransformations using cofactor-dependent enzymes will be described. In particular, biotransformations of commercial interest will be examined. A brief introduction to coenzyme characteristics and properties will also be given.

5.2 Coenzyme-dependent Reactions

Coenzymes work as electron acceptors during the enzymatic dehydrogenation of substrates. Figure 5.1 shows the structure of coenzymes containing nicotinamide as the main constituent: the nicotinamide–adenine dinucleotide (NAD), also called diphosphopyridine nucleotide (DNP), and the nicotinamide–adenine dinucleotide phosphate (NADP), also called triphosphopyridine nucleotide (TNP).

Four different types of oxidoreductases catalyze oxidation–reduction reactions that require the cooperation of coenzyme: pyridine nucleotide-dependent dehydrogenases, which require NAD or NADP as coenzyme; flavin-dependent dehydrogenase, which requires flavin–adenine dinucleotide (FAD) as prosthetic group; ferredoxins; and cytochromes.

The pyridine nucleotide-dependent dehydrogenases are the most widely used in industrial production. The catalytic mechanism of these systems is illustrated in Figure 5.2. The coenzyme interacts with coenzyme and substrate. A hydrogen ion is transferred from

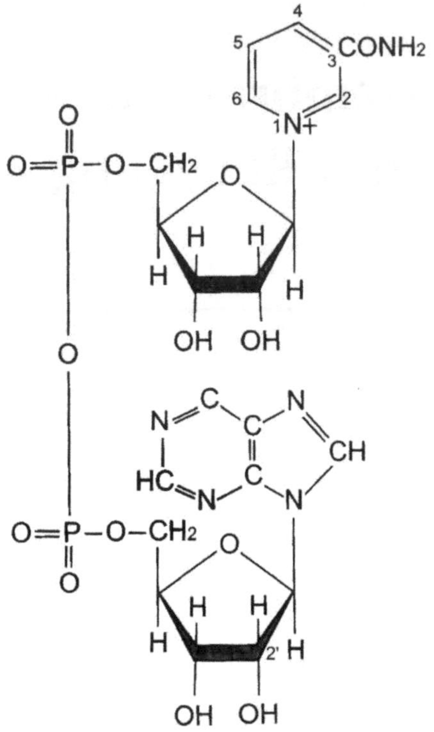

Figure 5.1 Structure of NAD cofactor.

Reduced substrate + NAD$^+$ ⇌ Oxidized substrate + NADH + H$^+$

(NAD$^+$)   (NADH)

Figure 5.2 General reactions catalyzed by piridine nucleotide-dependent hydrogenases.

the substrate to the NAD$^+$, giving the corresponding reduced coenzyme; the other hydrogen removed from the substrate is present in solution as hydrogen ion.

The pyridine nucleotide-dependent dehydrogenases show a Michaelis–Menten behaviour towards both substrate and cofactor. Most of the dehydrogenases studied show the kinetic behaviour of ordered reactions with double substrate: the cofactor interacts first with the enzyme; after the hydrogen has been transferred from the substrate to the cofactor, the oxidized substrate is released followed by the reduced cofactor.

NAD$^+$ and NADP$^+$ are present as open and closed shapes. The closed or overlapped shape (in which the nicotine amide and adenine rings are parallel) is more prevalent in

Table 5.1 Enzymatic systems requiring NAD or NADP as coenzyme

Enzyme	Reaction catalyzed
Methane monoxygenase from *Methyloccus capsulatus* and *Methylosinus tricosporium*	$CH_4 + O_2 + NAD(P)H + H^+ \rightarrow CH_3OH + H_2O + NAD(P)^+$
5α-reductase	androsterone + NADPH $\rightleftharpoons$ dihydrotestosterone + NADP$^+$
Glucose dehydrogenase from *Gluconobacter suboxydans*	β-D-glucose + NAD(P)$^+$ $\rightleftharpoons$ δ-gluconolactose + NAD(P)H
Lactate dehydrogenase	lactic acid + NAD$^+$ $\rightleftharpoons$ pyruvic acid + NADH
Malate dehydrogenase	malic acid + NAD$^+$ $\rightleftharpoons$ oxaloacetic acid + NADH
Leucine dehydrogenase	pyruvic acid + NH$_3^+$ + NADH + H$^+$ $\rightleftharpoons$ L-alanine + NAD$^+$
Glyceraldehyde-3-phosphate dehydrogenase	glyceraldehyde-3-phosphate + HPO$_4^{2-}$ + NAD$^+$ $\rightleftharpoons$ 1,3-disphoglycerate + NADH + H$^+$
Horse liver alcohol dehydrogenase	3-methylpentane-1,5-diol + NAD$^+$ $\rightleftharpoons$ (3S)-(−)-3-methylvalerolactone + NADH
Alcohol dehydrogenase	limonoate A-ring lactone + NADH$^+$ $\rightleftharpoons$ 17-dehydrolimonoate A-ring lactone + NADH
Methanol dehydrogenase	CH_3OH + methoxatin^a $\rightleftharpoons$ CH_2O

^{a} Methoxatin is a coenzyme.

solution; when NAD$^+$ interacts with the enzyme it assumes the open shape. The three-dimensional structures of some pyridine nucleotide-dependent dehydrogenases have been determined, such as those of lactate dehydrogenase, malate dehydrogenase, liver alcohol dehydrogenase and glyceraldehyde-3-phosphate dehydrogenase. In general, pyridine nucleotide-dependent dehydrogenases, are stereoselective for the substrate (for example, lactate dehydrogenase is stereoselective for L-lactic acid) and for the cofactor (they require NAD or NADP, and transfer the hydrogen only from one side of the pyridine ring). Only few enzymes, such as glutamic dehydrogenase, can work with both cofactors.

Table 5.1 lists enzymatic systems that require NAD or NADP.

All of these aspects of catalytic systems can greatly influence their performance and have to be taken into account in process design and development on large scale.

5.3 Catalytic Membrane Reactors Using Coenzyme-dependent Reactions

In reactor systems using coenzyme-dependent reactions, the membrane system is intended to retain both enzyme and coenzyme in the reaction site and to separate the reaction products.

The problems that must be faced in developing chemical reactors using coenzyme-dependent oxidoreductases include (1) regeneration of cofactor — as reaction proceeds the coenzyme is reduced (or correspondingly oxidized, depending on the reaction of interest) and in order to be reused it needs to be reconverted to the oxidized (or reduced) form; (2) retention of enzyme and coenzyme in the reaction vessel; and (3) separation of product from substrate and by-products, which often have similar molecular size and cannot be separated by traditional filtration processes.

5.3.1 *Regeneration of Coenzyme*

Regeneration and recycling of cofactor must be taken into account to obtain good efficiency of cofactor-dependent enzymatic systems. Regeneration of coenzyme can be achieved by coupling conjugate enzyme-catalyzed reactions that use the coenzyme in opposite directions. In this way, as a reaction of interest proceeds and the cofactor is, say, reduced, a conjugated enzyme that uses the reduced form is utilized to reconvert it to the oxidized form. Some of these enzymatic systems of commercial interest are summarized in Table 5.2.

Table 5.2 Conjugated reactions for regeneration of coenzyme

Conjugated reaction system	References
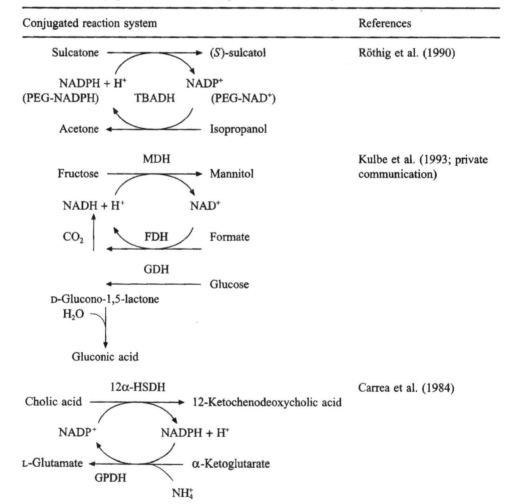	Röthig et al. (1990)
	Kulbe et al. (1993; private communication)
	Carrea et al. (1984)

Table 5.2 (cont'd)

Conjugated reaction system	References

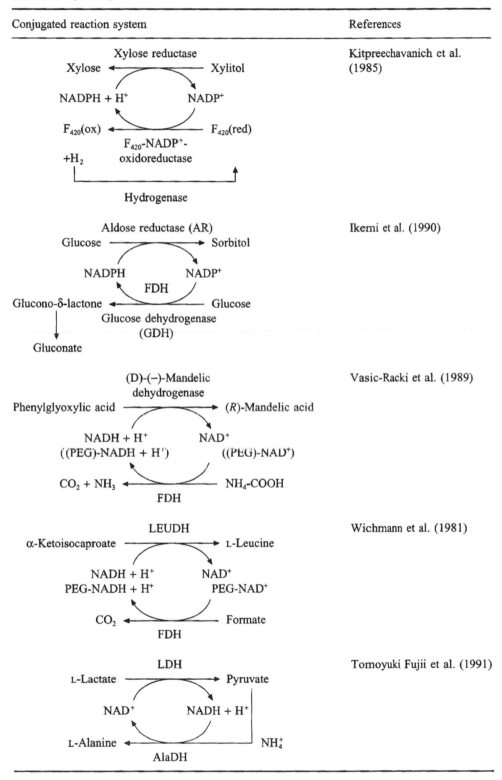

Kitpreechavanich et al. (1985)

Ikemi et al. (1990)

Vasic-Racki et al. (1989)

Wichmann et al. (1981)

Tomoyuki Fujii et al. (1991)

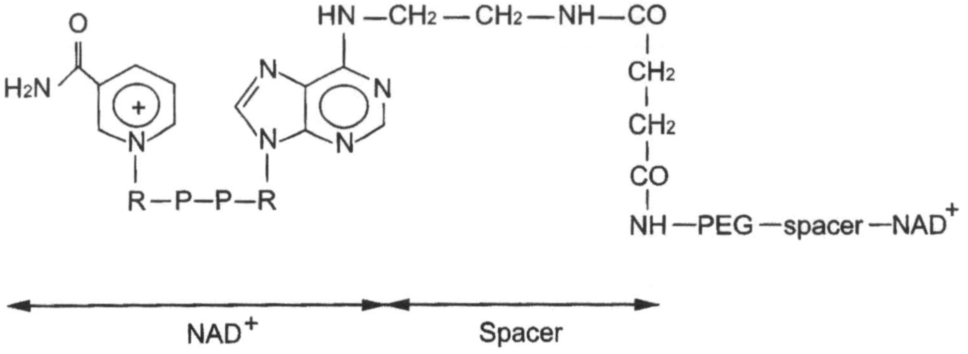

Figure 5.3 Structure of poly(ethylene glycol) bound NAD^+. R = ribose; P = phosphate; PEG = poly(ethylene glycol), $(CH_2\text{-}CH_2\text{-}O)_n$.

5.3.2 *Retention of Coenzyme*

In designing biocatalytic processes using cofactor-dependent enzymes, profitable applications on a long-term basis will probably depend upon the use of continuous processing with immobilized biocatalysts and with the necessary cofactor retained within the reactor. For higher production rate by continuous operation, enzyme and cofactor retention within the bioreactor becomes an economic necessity.

Various approaches for retaining the coenzyme have been explored. Since cofactor molecular masses are relatively small (e.g. the molar mass of NADH is $662\ g\ mol^{-1}$), their separation from substrates and product by filtration is difficult. One method used to retain the cofactor is to increase its molecular size by binding it to other molecules, such as water-soluble polymers (Wichmann et al., 1981; Bückmann et al., 1987) or enzymes (Mazid and Laidler, 1982). Poly(ethyleneimine) (PEI) and poly(ethylene glycol) (PEG) are soluble polymers widely used for these purposes. Poly(ethylene glycol) is suitable for these applications because it is a linear polymer with only two good and accessible terminal ends on the polymer chain, so that steric hindrance should be minimized. Its good solubility in water is also important. The structure of PEG-bound NAD^+ is reported in Figure 5.3. The polymer–cofactor complex can easily be retained by ultrafiltration membranes. This principle is not generally applicable, because some enzymes do not accept derivatized coenzymes and for many others the kinetic constants are changed drastically. Table 5.3 describes enzymes active with native or derivatized cofactors.

L-Leucine dehydrogenase with PEG-NAD(H) was used to produce L-leucine from α-ketoisocaproate (Wichmann et al., 1981). A conjugated reaction catalyzed by formate dehydrogenase was used to regenerate the cofactor. Ultrafiltration Amicon YM 5 membranes with cut-off of 5 kDa were used to retain the cofactor. The reactor had a volume of 10 ml; the substrate flow was $2\ ml\ h^{-1}$, resulting in a residence time of 5 h. L-Leucine was produced continuously for one month at an average conversion of about 98% (a maximal conversion of 99.7% was reached around the twentieth day), with a yield of $42.5\ g\ l^{-1}\ day^{-1}$.

The same type of membrane reactor was used to obtain continuous conversion of phenylglyoxylic acid to (R)-mandelic acid by mandelic-acid dehydrogenase, with simultaneous regeneration of PEG-NAD^+ by formate dehydrogenase (Vasic-Racki et al., 1989). In this case, a maximal conversion of 98% was achieved during the first 12 days at a residence time of 2 h; a yield of $700\ g\ l^{-1}\ day^{-1}$ was obtained.

Table 5.3 Enzymes active with derivatized coenzymes

Enzyme	Activity with coenzyme	Reference
L-Leucine dehydrogenase (LEUDH) and formate dehydrogenase (FDH)	Show the same kinetic constants with native or PEG-10000-NAD(H)	Wichmann et al. (1981)
Glucose dehydrogenase (GlDH)	Not active with PEG-NAD$^+$	Kula and Wandrey (1987)
Horse liver alcohol dehydrogenase (HLADH)	Three to four times less active with PEG-NAD$^+$ compared to native NAD$^+$	Vanhommerig et al. (1996)
Mandelic-acid dehydrogenase (MADH)	Active with PEG-NADH	Vasic-Racki et al. (1989)
Alcohol dehydrogenase from *T. brokii* (TBADH)	Active with PEG-NAD(H)	Röthig et al. (1990, personal communication)

As can be seen, derivatized coenzymes can have differing efficiencies depending on the type of enzymatic systems with which they must interact. The observed activities of cofactors complexed with dehydrogenases are not yet understood at the molecular level. The low functionality is probably caused by steric unavailability or by attachment to a position of the cofactor molecule that is necessary for its function. To minimize steric hindrance, the coenzyme can be co-attached to the soluble polymer with spacer arms of various lengths. The benefits of this vary with the type of spacer arm and also depending on the system investigated.

Vanhommerig et al. (1996) carried out kinetic and modelling studies, including molecular mechanics and dynamics, using PEG-NAD$^+$ and horse liver dehydrogenase (HLADH) as enzyme. In these studies they pointed out that the kinetic behaviour of PEG-NAD$^+$ was affected by interference of singly substituted poly(ethylene glycol) in the enzymatic reaction, steric hindrance exerted by the PEG tail, and destabilization of the HLADH dimer.

Because several enzymes do not interact efficiently with modified coenzymes, membrane technologies have been investigated with a view to retaining the coenzyme in the native form. Negatively charged ultrafiltration membranes can be used to retain nicotinamide–adenine dinucleotides in a reaction system. These coenzymes are amphoteric molecules with acidic phosphate and basic amino groups which at pH higher than 3 carry a negative net charge. Thanks to this property, they can be rejected by the electrostatic repulsion between the negatively charged groups of the ultrafiltration membrane and the coenzyme itself. In this way, during ultrafiltration with charged membrane, enzyme and coenzyme are retained in the reaction vessel, while uncharged small molecules (substrates, product, by-products, etc.) can permeate through the membrane (Figure 5.4). Companies producing charged membranes are listed in Table 5.4. When native coenzymes can be recycled a sufficient number of times, their cost is no longer a limiting factor for the overall process.

Catalytic membrane reactors can be implemented using charged ultrafiltration membranes in flat-sheet or tubular configuration. In one case, stirred tank reactors (STRs) with dead-end ultrafiltration mode can be obtained. In the other, plug-flow reactors with cross-flow ultrafiltration mode are possible.

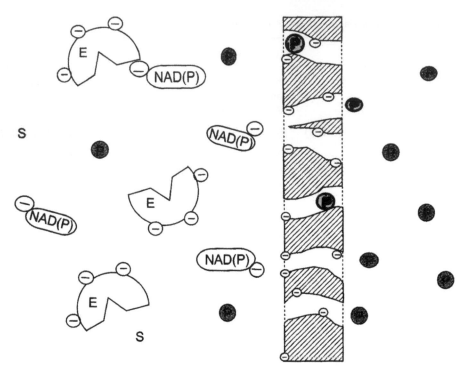

Figure 5.4 Principle of the charged ultrafiltration membrane reactor. E = enzyme; S = substrate; P = product.

Table 5.4 Companies producing charged membranes

Company	Membrane type
Nitto (Japan)	NTR 7400 (10/30/50)
Amicon (USA)	Y05
	YM5
	SP326
Desalination Systems (USA)	DS5
	G5
Osmonic (USA)	SEPA-MX07
TechSep (France)	Organo-mineral membranes

The capacity of the membrane to retain the coenzyme in the reactor volume can be determined experimentally by measuring the retention coefficient, defined as

$$R = 1 - \frac{C_p}{C_f}$$

where C_p and C_f are the coenzyme concentration in the permeate and initial feed solution, respectively.

Table 5.5 Retention coefficient of charged UF membrane towards NAD(P)H coenzyme

Type of membrane	Supplier	Cofactor	Rejection	Buffer solution and pH	Reference
NTR-7410	NITTO	NADPH	0.87	Tris-Cl buffer; (pH 7.5)	Ikemi et al. (1990)
NTR-7410	NITTO	NADH	0.73	Phosphate buffer (pH 6–7.5)	Nidetzky et al. (1994a,b)
NTR-7410	NITTO	NADP(H)	0.59	Phosphate buffer (pH ~8.0)	Giorno et al. (1993)
NTR-7410	NITTO	NADH	0.65	Glycine buffer (pH ~10)	Giorno et al. (1997)
NTR-7430	NITTO	NADH	0.99	Phosphate buffer (pH 6–7.5)	Nidetzky et al. (1994a,b)
NTR-7430	NITTO	NADPH	0.97	Glycylglycine buffer (pH 7.5)	Kitpreechavanich et al. (1985)
NTR-7450	NITTO	NADP(H)	0.95	Phosphate buffer (pH ~8.0)	Giorno et al. (1993)
Y05	AMICON	NADH	0.89	Phosphate buffer (pH 6–7.5)	Nidetzky et al. (1994a)
DS5	Desalination	NADH	0.98	Phosphate buffer (pH 6–7.5)	Nidetzky et al. (1994b)
PEEK+BSA	Lab-made	NADP(H)	0.65	Phosphate buffer (pH ~11)	Giorno et al. (1993)

Retention of coenzyme depends strongly upon the uniformity of membrane structure and charge density on the membrane surface. The ionic strength of solutions strongly influences the permeselectivity of charged ultrafiltration membranes. In particular, increasing ionic strength leads to a decrease of rejection of negatively charged molecules. Therefore, in such enzyme processes, low ionic strength may be maintained in the reaction system. The type of buffer solution and pH used also influences the permeselectivity and retention properties of charged membranes.

Table 5.5 reports the retention coefficients of charged membranes towards nicotinamide–adenine dinucleotide coenzymes. Flat sheet UF-charged membranes, such as NTR-7400 membrane series (produced by Nitto, Japan) are composite membranes, where the skin layer is made of sulfonated polyether sulfone, with a thickness of about 0.3 μm and a charge density of about 1.5 meq g^{-1}; the support layer is neutral polysulfone (Ikeda et al., 1988). NTR-7400 membranes have been used by several authors to realize UF-charged membrane reactors for retention of cofactors (Kitpreechavanich et al., 1985; Ikemi et al., 1990; Röthig et al., 1990; Giorno et al., 1993; Nidetzky et al., 1994).

In general, for the systems studied, a higher efficiency of catalytic charged membrane systems is observed than with batch systems and membrane reactors using derivatized cofactor.

The production of sulcatol by alcohol dehydrogenase from *Thermoanaerobium brockii* (TBADH) using native NADP(H) or PEG-NADP(H) was carried out by Röthig et al.

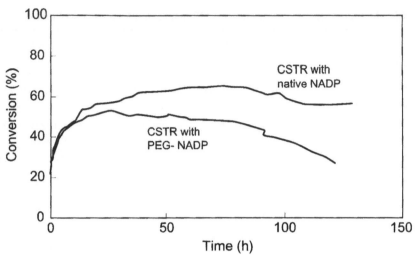

Figure 5.5 Comparison of native NADP to PEG-NADP in a CSTR equipped with an integrated UF-membrane (from Röthig et al., 1990).

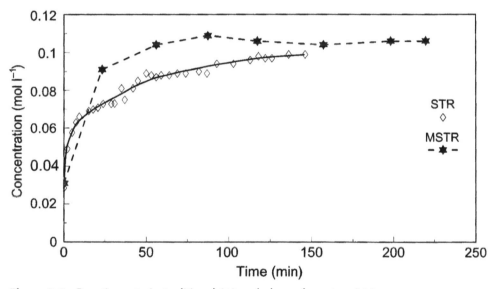

Figure 5.6 Reaction rate in traditional (◊) and charged reactors (✱).

(1990). TBADH showed a reduced V_{max} (84% of that with NADP(H)) and an increased K_m (33 µmol l^{-1}) with PEG-NADP(H). The process with native NADP(H) gave a conversion higher by 10% than a system with derivatized coenzyme (Figure 5.5).

Charged membrane reactors (in STR configuration) for conversion of lactate to L-alanine by lactate dehydrogenase and alanine dehydrogenase using NAD(H) coenzyme have been compared with batch traditional reactors (STR). Both systems used the coenzyme in the native form. The reactions were carried out at 25 °C, pH 10. NTR-7410 membranes were used for the experiments with the charged membrane reactor, and at an applied transmembrane pressure of 2 bar the permeate flux was about 10 l h^{-1} m^{-2}. As shown in Figure 5.6 the reaction rate of enzymatic systems in charged membrane reactors was higher than in the traditional system.

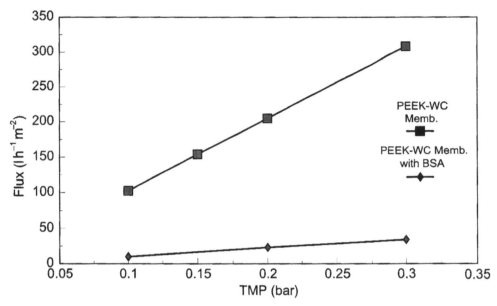

Figure 5.7 Permeate flux as a function of transmembrane pressure through PEEK-WC membranes with (■) and without BSA (♦).

NADP(H)-dependent aldose reductase from *Candida tropicalis* IAM 12202 was used to convert glucose into sorbitol in a charged membrane reactor with NTR-7410 membranes (Ikemi et al., 1990); a UHP-43 ultrafiltration apparatus (Advantec Toyo Kaisha Ltd) was used. The initial substrate concentration was 1.33 mol l^{-1}. The system operated for more than 33 days with a substrate conversion of 85% and a productivity of 114 g l^{-1} day^{-1}; a turnover number of 106 000 for NADP(H) was reached.

Charged membranes can be obtained by immobilizing a protein, such as bovine serum albumin (BSA) on the membrane surface. At basic pH the BSA is negatively charged and it gives the membrane the ability to reject the negatively charged coenzyme. Such composite membranes have been obtained by ultrafiltering BSA in phosphate buffer solution at pH 8 through a microporous polymeric membrane made of polyetheretherketone (PEEK). As a consequence of the concentration polarization phenomenon, the BSA gelified on the membrane surface. The evidence that protein is immobilized is given by mass balance of protein between initial and final solution, and by comparison of the permeate flux through the membrane before and after ultrafiltration of protein solution (Figure 5.7). The retention coefficient of these membranes towards NADP(H) increases linearly with increasing pH (Figure 5.8) (Giorno et al., 1993).

Theoretical descriptions of bioreactor systems with coenzyme regeneration have been attempted by several authors. Katayama et al. (1983) constructed a fairly complete model for the conjugated system of lactate dehydrogenase and alcohol dehydrogenase with NAD regeneration, considering the backward reactions involved by using a multisubstrate Michaelis–Menten expression for the enzyme kinetics and assuming a Theorell–Chance mechanism. In this work no mass transfer resistance was considered and the reactor was treated as a complete mixing vessel. This assumption could be made since polymer-bound coenzyme was compartmentalized together with the conjugated enzyme in the ultrafiltration membrane reactor to which substrate was fed continuously and from which product was removed through the membrane.

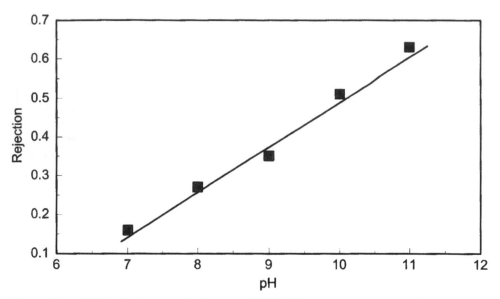

Figure 5.8 NADP rejection vs pH through a PEEK-WC membrane functionalized with BSA.

Ishikawa et al. (1989) constructed a model for a hollow-fibre reactor considering the concentration distribution in the axial direction and assuming plug flow. For the enzyme kinetics, a multisubstrate Michaelis–Menten form was used with the assumption of a Theorell–Chance mechanism.

Tomoyuki Fujii et al. (1991) constructed a theoretical reactor model for the production of L-alanine with coenzyme regeneration. Considering the concentration distribution and mixing in the axial direction in a hollow-fibre capillary reactor, performance of the reactor was theoretically analyzed with a multistage stirred tank reactor model combined with the kinetic model based on all the elementary reactions involved.

5.3.3   *Separation of Product*

Scale-up of coenzyme-dependent reaction sequences to technical processes requires retention of enzymes and coenzymes as well as separation from the product streams. The whole process, even when constituted by different integrated steps, can be seen as a catalytic membrane reactor that allows chemical reaction of interest to be carried out and simultaneous separation of product.

Kulbe et al. (1993; private communication) developed an integrated system using ultrafiltration and electrodialysis for production of mannitol from fructose, regenerating the cofactor by the conjugated conversion of glucose to gluconic acid (Table 5.2). As shown in Figure 5.9 the integrated system comprises an ultrafiltration step, downstream of the enzyme reactor, to retain enzymes. The permeate containing coenzymes and products is fed to an electrodialysis system using bipolar membranes, which separates charged products such as gluconic acid while the coenzyme cannot pass the anion exchange membrane because of its size. The uncharged product mannitol is removed through a charged

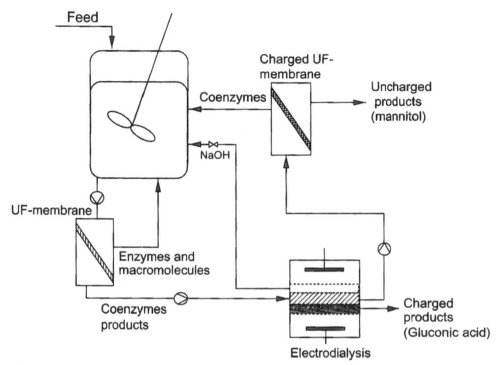

Figure 5.9 Continuous production of mannitol and gluconic acid and rejection of enzymes and coenzymes.

ultrafiltration membrane (after the electrodialysis step), while the coenzyme is retained in the reactor by electrostatic repulsion.

In the reaction system L-lactate → pyruvate → L-alanine, the coenzyme retention was achieved using a charged ultrafiltration membrane; a retention coefficient of about 0.70 was obtained. Furthermore, the L-alanine has to be separated from other components, such as lactate and pyruvate. The use of membrane-based solvent extraction through a charged membrane after the ultrafiltration charged system is currently under investigation.

5.4 Conclusions

The real potential of membrane reactors becomes evident in conjunction with coenzyme-dependent enzymatic systems. Coenzymes, like NAD or NADP, usually have a long-term effect on enzyme activity only if they can move from one enzyme, able to oxidize them, to another, able to reduce them, in loop kinetics. Continuous homogeneous catalysis is a prerequisite for achieving high reaction yields. Enzyme membrane reactors offer a suitable reaction environment, are able to retain the coenzymes in the reaction system, and can be integrated with other membrane processes for the separation and purification of products.

So far, the technology has been developed at laboratory scale; nevertheless, encouraging results suggest that continuous coenzyme-dependent processes should no longer be limited by coenzyme costs.

5.5   References

BÜCKMANN, A.F., MORR, M. and KULA, M.-R., 1987, Preparation of technical grade poly-ethylenglycol (PEG) (MR = 20 000)-N^6-(2-aminoethyl)-NADH by a procedure adaptable to large-scale synthesis, *Biotechnol. Appl. Biochem.*, **9**, 258–268.

CARREA, G., BOVORA, R., CREMONES, P. and LODI, R., 1984, *Biotechnol. Bioeng.*, **26**, 560–563.

GIORNO, L., MOLINARI, R., DRIOLI, E., BIANCHI, D. and CESTI, P., 1993, Use of charged UF-membrane bioreactors for retention of cofactors, *Sixth European Congress on Biotechnology*, Firenze.

GIORNO, L., RAIMONDI, G. and DRIOLI, E., 1997, Charged membrane reactor for coenzyme dependent dehydrogenation reactions, in *Catalysis on the Eve of the XXI Century. Science and Engineering*, pp. 75–76, Memorial G.K. Boreskov Conference, Novosibirsk.

IKEDA, K., NAKANO, T., ITO, H., KUBOTA, T. and YAMAMOTO, S., 1988, New composite charged reverse osmosis membrane, *Desalination*, **68**(109), 389–390.

IKEMI, M., KOIZUMI, N. and ISHIMATSU, Y., 1990, Sorbitol production in charged membrane bioreactor with coenzyme regeneration system: I. selective retainment of NADP(H) in a continuous reaction, *Biotechnol. Bioeng.*, **36**, 149–154.

ISHIKAWA, H., TANAKA, T., TAKASE, S. and HIKITA, H., 1989, *Biotechnol. Bioeng.*, **34**, 357.

KATAYAMA, N., URABE, I. and OKADA, H., 1983, *Eur. J. Biochem.*, **132**, 403–407.

KITPREECHAVANICH, V., NISHIO, N., HAYASHI, M. and NAGAI, S., 1985, Regeneration and retention of NADP(H) for xylitol production in an ionized membrane reactor, *Biotechnol. Lett.*, **7**(9), 657–662.

KULA, M.R. and WANDREY, C., 1987, *Methods Enzymol.*, **136**, 9–21.

MAZID, M.A. and LAIDLER, K.J., 1982, Kinetics of yeast alcohol dehydrogenase and its coenzyme co-immobilized in a tubular flow reactor, *Biotechnol. Bioeng.*, **24**, 2087–2097.

NIDETZKY, B., NEUHAUSER, W., HALTRICH, D. and KULBE, K.D., 1994a, Continuous re-generation of NAD(H) by coupled aldose reductase and glucose dehydrogenase in a charged ultrafiltration membrane reactor, in *ICHEME — Applied Biocatalysis*, pp. 14–16, ICHEME, Warwickshire, UK.

NIDETZKY, B., SCHMIDT, K., NEUHAUSER, W., HALTRICH, D. and KULBE, K.D., 1994b, Application of charged ultrafiltration membranes in continuous enzyme-catalyzed processes with coenzyme regeneration, in PYLE, D.L. (ed.) *Separation for Biotechnology 3*, pp. 351–357, Royal Society of Chemistry, London.

RÖTHIG, T.R., KULBE, K.D., BÜCKMANN, F. and CARREA, G., 1990, Continuous coenzyme dependent stereoselective synthesis of sulcatol by alcohol dehydrogenase, *Biotechnol. Lett.*, **12**(5), 353–356.

TOMOYUKI FUJII, OSATO MIYAWAKI, and TOSHIMASA YANO, 1991, Modelling of hollow fibre capillary reactor for the production of L-alanine with coenzyme regeneration, *Biotechnol. Bioeng.*, **38**, 1116–1172.

VANHOMMERIG, S.A.M., SLUYTERMAN, L.A.Æ. and MEIJER, E.M., 1996, Kinetic and modelling studies of NAD$^+$ and poly(ethylene glycol)-bound NAD$^+$ in horse liver alcohol dehydrogenase, *Biochimi. Biophys. Acta*, **1295**, 125–138.

VASIC-RACKI, D., JONAS, M., WANDREY, C., HUMMEL, W. and KULA, M.-R., 1989, Continu-ous (*R*)-mandelic acid production in an enzyme membrane reactor, *Appl. Microbiol. Biotechnol.*, **31**, 215–222.

WICHMANN, R., WANDREY, C., BUCKMANN, A.F. and KULA, M.R., 1981, Continuous enzymatic transformation in an enzyme membrane reactor with simultaneous NAD(H) regeneration, *Biotechnol. Bioeng.*, **23**, 2789–2802.

6

Catalytic Membrane Reactors for Bioconversion of Low Water Solubility Substrates

6.1 Introduction

Besides the applications described in previous chapters, enzyme-catalyzed reactions are also used to convert substrates that have low solubility in water. The conversion of these substrates is carried out in multiphase systems where at least two phases are simultaneously present: aqueous phase (containing the biocatalyst) and organic phase (containing the substrate). Often, for these systems the enzyme is used in the immobilized form. In these cases, three phases are present: organic phase, aqueous phase and immobilized enzyme. The enzyme can be immobilized on solid microporous particles and dispersed in organic/water (O/W) emulsion, or immobilized on artificial membranes. As well as other enzymes, lipases have been widely used to catalyze conversion of low water solubility substrates. Often, lipases show enantiospecificity for one of the two enantiomers of a racemic substrate. In these cases, they can be used for enantiomeric resolution, as already discussed in Chapter 4.

The particular physicochemical properties of enzyme-catalyzed reactions in multiphase systems have promoted the development of new catalytic membrane reactor configurations, such as biphasic organic/aqueous membrane reactors. The capacity of enzymes to convert nonphysiological substrates in nonaqueous phases has also promoted the development of catalytic membrane reactors using pure organic solvents.

In the present chapter, the use of catalytic membrane reactors in biphasic organic/aqueous systems and in pure organic phase will be discussed.

6.2 Biphasic Organic/Aqueous Catalytic Membrane Reactors

Biphasic enzyme membrane reactors can be realized as emulsion enzyme membrane reactors (E-EMR) or as two-separated-phase enzyme membrane reactors (TSP-EMR).

In the first case, the biocatalyst (enzyme) is freely suspended in the emulsion and retained in the reaction mixture by the membrane. The phase that contains the product, usually the aqueous phase, is separated through ultrafiltration membranes. The reaction volume is kept constant by continuous addition of new aqueous phase as it permeates

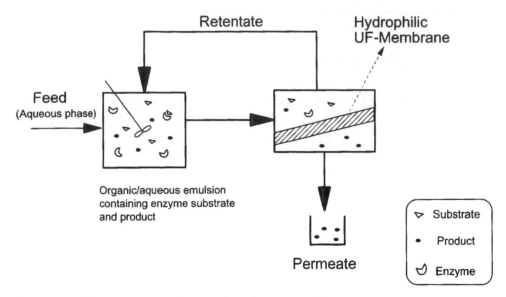

Figure 6.1 Schematic representation of emulsion organic/aqueous enzyme membrane reactor.

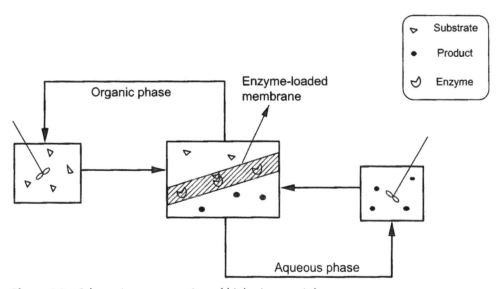

Figure 6.2 Schematic representation of biphasic organic/aqueous enzyme membrane reactor.

through the membrane. The case where the reactor product is soluble in the aqueous phase and a hydrophilic ultrafiltration membrane is used to remove the aqueous phase is illustrated in Figure 6.1. Provided the enzyme is not affected by shear-friction, this kind of system has the advantage of working at high reaction rate, but it has the limitation that the product is present in a diluted form.

In biphasic enzyme membrane reactors with two separated phases, the enzyme is immobilized in or on the membrane and the enzyme-loaded membrane separates the two immiscible phases (Figure 6.2). The substrate is present in the organic phase which

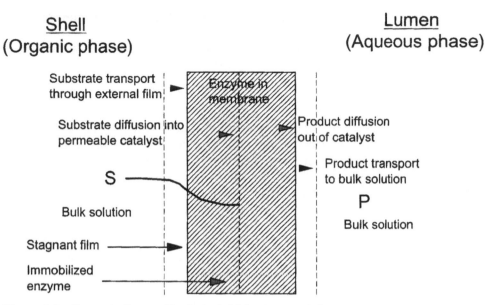

Figure 6.3 Concentration profile through biphasic organic/aqueous enzyme membrane reactor.

diffuses through the membrane carrying the substrate to the enzyme, where the reaction of the substrate takes place; the product, which has high water solubility, is extracted in the aqueous phase. Of particular interest is the case where the enzyme is enantiospecific and converts only one of the isomers of the substrate, giving the production and separation of enantiomeric compounds in one step (see Chapter 4, Figure 4.6). The concentration profile through a two-separated-phase enzyme-loaded membrane is illustrated in Figure 6.3.

In biphasic organic/aqueous membrane reactors with the two liquid phases separated, the membrane contains the enzyme and acts as a catalytic interface; on one side it is in contact with the organic phase and on the other with the aqueous phase. When the enzyme is immobilized by physical methods, it can contain the enzyme reversibly, and this allows removal of the inactivated catalyst and reuse of the membrane as support for new catalyst.

The two phases are kept in contact at the level of the membrane pores. The use of tubular membranes provides a high exchange surface per unit volume, with the general result of increasing transport rate. The fact that the organic/aqueous interface is located at the membrane level means that the exchange surface is well defined, independently of the level of stirring and volume of phases. This system is generally more flexible than emulsion reactors and is easier to scale up.

Although the two phases are kept in contact, they are also prevented from dispersing into each other by the membrane itself. When hydrophilic membranes are used, the aqueous phase wets the membrane, penetrating into the porous structure. At the interface with the organic phase, the aqueous phase is maintained in the membrane interface by applying a higher transmembrane pressure from the organic to the aqueous side. The parameters that dictate the order of magnitude of transmembrane pressure are the mechanical resistances of membrane fibres and the breakthrough pressure of the membrane.

6.2.1 *Parameters That Influence Selection of Reactor Components*

The components of a biphasic organic/aqueous membrane reactor with the two phases separated comprise the organic phase, the aqueous phase, the membrane, and the enzyme.

Organic phase

The organic phase contains the substrate. Often, the substrate itself is the organic phase, if it is immiscible with water and has low viscosity. Otherwise, it is diluted in an organic solvent that is immiscible with water. The use of pure substrate as organic phase is advantageous since the mass transfer resistances are minimized because no significant concentration gradients are present. In the case of enzymatic resolution of stereoisomers, the unconverted isomer can be considered as the solvent, producing concentration gradients of the isomer removed.

The choice of organic solvent to contain the substrate should be made on the basis of substrate solubility, solvent viscosity, influence of solvent on enzyme activity and stability, membrane resistance to solvent, and miscibility of solvent with water.

To minimize the amount of solvent used to dissolve the solute, the solvent must show high affinity for the substrate; in this way, reduced concentration gradients can be obtained and a high repartition of substrate in the aqueous phase (within the membrane containing the enzyme) can also be reached. A low viscosity of solvent increases the transport rate of solute (substrate) to the enzyme, increasing reaction rate; the solvent must be immiscible with water in order to prevent mixing of the phases.

Aqueous phase

The aqueous phase is constituted by a buffer solution; it is chosen on the basis of optimal pH value and ionic strength for the enzyme. Additionally, when hydrolysis reactions are studied, the effect of repartition of the acid produced must be considered.

The partition coefficient of organic acids is defined as the concentration of undissociated acid [HA] in the organic phase divided by the concentrations of the acid and its anion [A⁻] in the aqueous phase:

$$K_{app} = \frac{[HA]_{org}}{[A^-]_{aq} + [HA]_{aq}}$$

The degree of dissociation of an acid is a function of the pH at which the reaction is carried out, and it must be considered together with the optimal value for enzyme activity.

The buffer concentration is also a parameter to be optimized. Reactions that produce acids create a pH gradient in the environment containing the enzyme. If the enzyme activity is highly susceptible to changes in pH, the use of high buffer concentration reduces the gradient.

Membranes

When hydrophobic microporous membranes are used to implement a biphasic organic/aqueous membrane reactor, the organic phase will wet the membrane and will diffuse through the pores to the other side of the membrane (Figure 6.4). The organic/aqueous interface will be immobilized at the membrane surface from the aqueous side by use of a transmembrane pressure equal to or higher than that of the organic phase.

Hydrophobic membrane

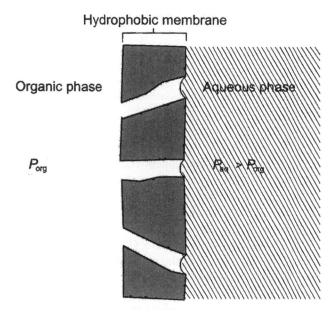

Figure 6.4 Organic aqueous interface in a biphasic membrane system using a hydrophobic membrane. (P = pressure.)

On the other hand, when hydrophilic microporous membranes are used, the organic/aqueous interface is located at the surface of the membrane on the side of the organic phase.

In general, the kinetic behaviour of these reaction systems cannot be interpreted by a traditional Michaelis–Menten model because the observed reaction rate is strongly influenced and limited by mass transfer, a parameter that is not taken into account in the Michaelis–Menten equation. The theory for membrane-based solvent extraction has been developed primarily for thin, uncharged symmetric membranes with no variation in porosity and pore size across the membrane thickness. It has been described in detail by Prasad and Sirkar (1992).

The solute flux from an aqueous to an organic phase (separated by a hydrophobic membrane) can be written in terms of individual mass transfer coefficients as

$$J_{Ni} = k_{iw}(C_{iwb} - C_{iw})$$
$$= k_{imo}(C'_{io} - C''_{io})$$
$$= k_{io}(C''_{io} - C''_{iob})$$

where

J_{Ni} = molar flux of species i [mol L^{-2} T^{-1}]

k_{iw} = mass transfer coefficient of species i in aqueous phase [L T^{-1})

C_{iwb} = concentration of component i in the bulk aqueous phase [mol L^{-3}]

C_{iw} = concentration of component i in the aqueous phase at a membrane interface [mol L^{-3}]

k_{imo} = mass transfer coefficient of species i in the organic phase present in the membrane [L T^{-1}]

C'_{io} = concentration of component i in the organic phase at the membrane interface [mol L^{-3}]

C''_{io} = concentration of component i in the organic phase present in the membrane [mol L^{-3}]

k_{io} = mass transfer coefficient of component i in the organic phase

C''_{iob} = concentration of component i in the bulk of organic phase.

Considering an overall mass transfer coefficient for the organic phase (K_o), the following relation can be used:

$$J_{Ni} = K_o(C'_{io} - C''_{iob})$$

where C'_{io} is the concentration of component i in the organic phase at the membrane interface in equilibrium with the concentration of component i in the aqueous phase.

Similar relations can be derived for a biphasic system separated by a hydrophilic membrane, with aqueous phase in the membrane pores and with extraction occurring from the aqueous to the organic phase.

Kiam et al. (1984) and Prasad and Bhave (1986) have demonstrated that the mass transfer coefficient is not influenced by the value of transmembrane pressure, unless this is approaching the breakthrough pressure. The mass transfer coefficient is influenced by the axial flow rate of phases.

The selection of membrane must consider performance in terms of reaction rate, mass transfer and stability of the biphasic system. This last point means that the two phases must be maintained in contact but separated at the level of pores (on the membrane surface or inside the membrane itself).

It has been previously stated that the organic/aqueous interface is kept immobilized by a higher pressure from the side of the non-wetting phase. It is important not to exceed the value of the breakthrough pressure in order to avoid mixing of the phases. Clearly the stability of biphasic organic/aqueous membrane systems can be achieved more easily if membranes with high limiting value of breakthrough pressure are used.

When cylindrical pores are considered, breakthrough pressure is defined by the Laplace–Young equation:

$$\Delta P_b = \frac{2\sigma_{wn}}{r} \cos \theta_{wm} \tag{6.1}$$

where

r = pore radius at the line of contact between the three phases (m)

σ_{wn} = interfacial tension at the liquid–liquid interface (N m^{-1})

θ_{wm} = angle of contact between wetting liquid and membrane.

The importance of these parameters in the selection of membranes for biphasic systems has been discussed by Vaidya et al. (1992).

As described in equation (6.1), the breakthrough pressure is inversely proportional to the pore radius. The maintenance of a stable liquid–liquid interface is therefore better achieved using a membrane with smaller pores (e.g. an ultrafiltration membrane) than using one with larger pores (e.g. microfiltration). Furthermore, it is advisable to use a liquid with high interfacial tension since breakthrough pressure is directly proportional to this parameter.

The contact angle between liquid phase and the pore entrance also has great influence on the maintanance of two separated liquid phases. Vaidya reported a detailed description of the influence on the breakthrough pressure of different pore geometries and contact angles between the wetting liquid and the membrane.

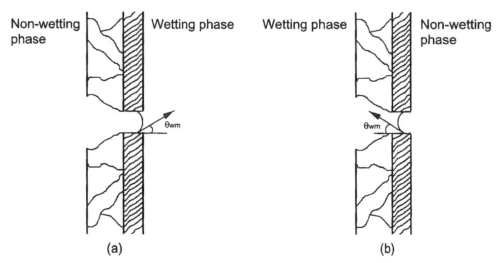

Figure 6.5 Contact angle between wetting liquid and membrane. (a) Non-wetting liquid on the open side of the pores. (b) Non-wetting liquid on the closed side of the pores.

In general, the configuration in which the non-wetting liquid is positioned on the open side of the pore (Figure 6.5a) is more stable than the opposite case (Figure 6.5b). However, the most stable configuration must be verified experimentally.

It should be taken into account that breakthrough pressure could change as the reaction proceeds. In fact, the stability of a liquid–liquid interface can be influenced by substances produced during the reaction. Products and by-products can change the interfacial tension, the contact angle, and so on, thus affecting the breakthrough pressure and the liquid–liquid interface stability.

In addition, the selection of membranes for two-separated-phase liquid reactors must be done on the basis of membrane resistance to organic solvents; membrane material (which affects the wetting characteristic, the contact angle between wetting phase and membrane, and the stability of immobilized enzyme); and membrane structure (pore size and geometry, pore size distribution, asymmetric or symmetric structure).

The stability of membranes to organic solvents is obviously an important factor affecting the performance and feasibility of biphasic systems. Polymeric membranes can be solubilized by organic solvents with damage to the membrane structure and permeselectivity properties.

Manufacturers generally provide information about the resistance of membranes to organic solvents. Where no data are available, the effect of organic solvents must be determined experimentally. Various polymeric membranes have been used for biphasic organic/aqueous systems; however, inorganic membranes are more advisable for reaction in pure organic phase.

A list of polymeric membranes used in biphasic systems and their resistance to organic solvents is given in Table 6.1.

The membrane material determines the hydrophobicity or hydrophilicity, which influences the wettability towards solvents. Hydrophilic membranes (such as polyamide, cellulose acetate, polyacrylonitrile) are preferentially wetted by the aqueous phase, whilst hydrophobic membranes (such as polypropylene, poly(vinylidene fluoride), polysulfone) are preferentially wetted by the organic phase. Very often, membranes are prepared with modified polymers (PS, PVDF), giving membranes that show intermediate characteristics.

Table 6.1 Resistance of polymer membranes to organic solvents[a]

Solvent	Polyacrylonitrile (PAN)	Polyamide (PA)	Cellulose acetate (CA)	Polyimide (PI)	Polysulfone (PS)	Poly(vinyl fluoride) (PVDF)	Polypropylene (PP)	Modified polyetheretherketone (PEEK-WC)
Methanol	SC		SC	LC	LC	LC	LC	LC
Ethanol	SC	SC	–	LC	LC	LC	LC	LC
Propanol	LC	LC	LC	LC	LC	LC	LC	LC
Butanol	LC	LC	–				SC	–
Amyl alcohol	LC	LC	LC	–	LC	LC	LC	
Octanol	LC		–					–
Hexane	–	LC	LC		LC	LC	NC	LC
Cyclohexane	LC	LC	–					LC
Heptane	LC	LC	–					LC
Benzene	–		LC	LC	NC	LC	SC	NC
Toluene	LC		LC	NC	NC	LC	LC	LC
Xylene	LC		LC	–	NC	LC	SC	
Ethyl acetate	NC		NC	–	NC	SC	LC	
Vegetable oils	LC	LC	–					LC
Silicone oils	LC		–					
Acetone	SC	LC	NC		NC	NC	LC	NC
Dioxane	–	NC	NC	NC	NC	NC	LC	
Tetrahydrofuran	–	NC	NC	NC	NC	LC	SC	
Ethylene glycol	–		SC					
Ethyl ether	–	LC	LC	–	LC	LC	LC	
Carbon tetrachloride	NC		SC	–	NC	LC		LC
Chloroform	NC		NC	–	NC	LC	–	NC
Formaldehyde 37%	–		LC	–	LC	LC	LC	
Dimethylformamide	NC	NC	–					NC
Acetonitrile	SC	SC						

[a] LC = long-term compatibility; SC = short-term compatibility; NC = not compatible.

The type of liquid that preferentially wets such membranes in a two-phase reactor is not readily apparent.

Besides the capacity to stably maintain two separated liquid phases, membrane material also affects the catalytic activity and stability of the immobilized biocatalyst. Lipases, for example, are activated when adsorbed onto hydrophobic supports (Horiuti and Imamura, 1978).

To obtain a high mass transfer coefficient, the selection of the type of membrane can be made on the basis of the repartition coefficient, defined as

$$m_i = \frac{C_{io}}{C_{iw}}$$

where C_{io} and C_{iw} are the equilibrium concentrations of component i in the organic and aqueous phase, respectively. When m_i is greater than 1, the solute strongly prefers the organic phase. In these cases it is better to use hydrophobic membranes, with organic phase in the pores, and to extract the solute from the aqueous to the organic phase. If m_i is less than 1, the solute is more soluble in the aqueous phase. In these cases it is better to use a hydrophilic membrane, with aqueous phase in the pores, and to extract solutes from the organic to the aqueous phase.

Asymmetric membranes with small, cylindrical pores on the thin selective layer allow stable maintenance of two separated phases.

Enzyme

The type of enzyme is dictated by the biotransformation to be carried out. Lipases have been widely used in multiphase reactions thanks to their ability to act at the organic/water interface with different substrates and to their resistance to organic solvents.

Lipases hydrolyze esters that are insoluble in water. They act at the interface between lipid and water (Verger, 1980). In these systems, the interfacial area affects the reaction rate more than the substrate concentration in the bulk. Free fatty acids and n-alcohols inhibit lipase activity; thus continuous separation of reaction product can significantly increase productivity of reactors.

Although lipases are distinguished from esterases by their ability to hydrolyze water-insoluble esters, they do exhibit some activity towards substrate monomers in solution. The first step of the reaction, which precedes the formation of the enzyme–substrate complex, is adsorption of the lipase onto the surface of the substrate. The subsequent steps of the catalytic reaction proceed at the interface and ultimately the enzyme is regenerated with the release of product. Catalytic reactions are accelerated by adsorption of lipase to the interface that contains the substrate (Momsen and Brockman, 1981). The adsorption to the interface is nonspecific, but it is necessary to activate the enzyme. This explains why lipases are more active on water-insoluble substrates than on water-soluble esters.

Owing to these peculiar characteristics, the lipases are feasible for use immobilized at the interface of biphasic organic/aqueous membrane reactors. Depending on the properties of the reaction system, the lipase can be adsorbed onto hydrophobic membranes, entrapped within the porous structure of asymmetric membranes, or chemically bound to functional groups of the membrane supports.

The use of hydrophilic or hydrophobic membranes dictates the use of higher trans-membrane pressure from organic-to-aqueous phase or from aqueous-to-organic phase, respectively; the transmembrane pressure must be higher from the phase which does not wet the membrane to the other phase. Obviously, to facilitate contact between substrate and catalyst, the organic phase containing the substrate must always be on the side of the membrane containing the enzyme. These catalytic systems are strongly influenced by mass transfer of organic substrate through the membrane to the catalyst and of product from catalyst to the aqueous phase.

Parameters such as viscosity, membrane thickness and catalyst layer must also be considered in the design of these kind of reactors.

High concentrations of glycerides and other organic substrates lead to phase separation and simultaneously to the formation of an oil/water interface. The observed reaction rates of lipase-catalyzed reactions are strongly influenced by the available interfacial area. Theoretical interpretations of the activation of lipases by interfaces have been attempted by a number of authors. Some assume that the substrate is activated by the presence of an oil/water interface; others that the lipase undergoes a change to an activated form upon contact with an oil/water interface. Explanations for the first type of activation involve higher concentrations of substrates in the vicinity of the interface than in the bulk of the oil, and more suitable conformations or orientations of the lipid molecules for chemical reaction. Explanations for the latter type of activation involve the existence of separate adsorption sites and catalytic sites for the lipase; in this way, the lipase becomes catalytically active only after binding to the interface, or after a conformational change of the lipase upon approaching the oil/water interface.

Lipases are generally soluble in water. Most animal lipases exhibit pH optima in the alkaline region (about 8–9). However, depending upon the substrate used, the presence of salts, and the kinds of emulsifiers present, the optimum may be shifted to the acidic range. Most microbial lipases display maximum activity at pH values ranging from 5.6 to 8.5 and maximum stability in the neutral pH range. With respect to temperature, most lipases are optimally active between 30 and 40 °C. The thermostability of lipases varies considerably according to their origin: animal and plant lipases are usually less thermostable than microbial extracellular lipases. Lipase from a thermophilic strain of *Pseudomonas* is stable at 100 °C. The heat stability of lipases depends on whether substrate is present, probably because substrate removes excess water from the immediate vicinity of the enzyme and thus restricts its overall conformational mobility.

Lipases are strongly adsorbed at air/water interfaces. Lipase from *Candida cylindracea* was easily inactivated at an air/water interface owing to the high interfacial tension; this effect is enhanced by shear forces and increases with temperature. Lipase from *Candida cylindracea* has been reported to lose activity as a function of shearing time and shear rate.

Depending on the type of lipase and reaction conditions, lipases show varying specificities, summarized in Table 6.2.

The activities of lipases relative to various triglycerides change with temperature: a higher temperature favours the release of long-chain fatty acids over to the corresponding short-chain acids.

Very often, the use of mixtures of lipases takes advantage of different lipase specificities and makes possible the synergistic hydrolysis of substrates; for example, soy bean oil has been hydrolyzed using a mixture of lipases from *Penicillium* sp. and *Rhizopus nivens*.

Some of the most common inhibitors and promoters of lipase activity are summarized in Table 6.3.

Table 6.2 Lipase specificities

Lipase source	Substrate	Specificity	Reference
Penicillum cyclopium	Monoglycerides	Lipid class, positional	Okumura et al. (1980)
Candida (cylindracea-rugosa)	Mono-, di-, triglicerides	Lipid class (monospecific)	Benzonana and Esposito (1971)
Chromobacterium viscosum	Mono-, di-, triglycerides	Lipid class (monospecific)	Sugiura and Isobe (1975)
Pseudomonas fluorescens	Mono-, di-, triglycerides	Lipid class (monospecific)	Sugiura et al. (1977)
Mucor mieher	Triglycerides	Lipid class, 1,3 position (external bonds)	Sonnet (1988)
Rhizopus arrhizus	Triglycerides	Lipid class, 1,3 position (external bonds)	Sonnet (1988)
Aspergillum niger	Triglycerides	Lipid class, 1,3 position (external bonds)	Tsujisaka et al. (1977)
Geotrichum candidum	Triglycerides	Lipid class (specific for 2-position)	Jensen (1990)
	Acetonide	Stereochemical	Sonnet (1987)
	Cyclohexanol	Stereochemical	Langrand et al. (1985)
	Arylpropionic esters	Stereochemical	Melonville et al. (1990)
	Sugar alcohols	Stereochemical	Chopineau et al. (1988)
Saiken	Triglycerides and fatty acids	1,3 positional	Basheer et al. (1995a, 1995b)

Table 6.3 Lipase inhibitors and promoters

Lipase source	Inhibitor	Promoters	Reference
Aspergillum wentii		Sodium ions Calcium ions	Desnelle (1972)
Aspergillum niger	Sodium ions Calcium ion Ferric ions	Ferrous ions	Garcia et al. (1991)
Rhizopus archizus	Bile salts		Canioni et al. (1978)
Pseudomonas, Chromobacterium, Streptococcus	Ferric ions Ferric ions Ferric ions Fatty acids Alcohols	Bile salts Bile salts Bile salts	Adams and Brawley (1981) Sugiura and Isobe (1974) Sugiura et al. (1974) Rastrup-Nielsen et al. (1990)
Candida rugosa	Ferrous ions Mercury ions		Kang and Rhee (1989)

Hydrolysis

$$R^1COOR^2 + H_2O \rightleftharpoons R^1COOH + R^2OH$$

Ester synthesis

$$R^1COOH + R^2OH \rightleftharpoons R^1COOR^2 + H_2O$$

Alcoholysis

$$R^1COOR^2 + H_2O \longrightarrow R^1COOH + R^3OH \rightleftharpoons R^1COOR^3 + H_2O$$
$$R^2OH$$

Transesterification

$$R^1COOR^2 + R^3OH \rightleftharpoons R^1COOR^3 + R^2OH$$

$$R^1COOR^2 + R^3COOH \rightleftharpoons R^3COOR^2 + R^1COOH$$

Interesterification

$$R^1COOR^2 + R^3COOR^4 \rightleftharpoons R^1COOR^2 + R^3COOR^2$$

Figure 6.6   Reactions catalyzed by lipases and esterases.

Fatty acids and alcohols tend to inhibit lipase-catalyzed hydrolysis reactions. Fatty acids are thought to accumulate at the lipid/water interface, thereby modifying the surface tensions and blocking access of the enzyme to unreacted triglyceride molecules. Alcohols are thought to disrupt the three-dimensional architecture of the lipase.

Metal cations influence the catalytic activity of lipase. The presence of calcium ions usually increases the reaction rate. Sodium ions have a positive effect on some lipases (such as lipase from *Aspergillus wentii*) but partially inhibit lipases from *Aspergillus niger*. The activity of some lipases is enhanced by bile salts that act as emulsifying agents; in presence of fatty acids, they form soluble micelles that accelerate the diffusion of products away from the organic/aqueous interface. The positive effect of metal ions could be due to the formation of complexes with ionized fatty acids that change their solubilities and behaviour at interfaces, whereas negative effect can be attributed to competitive binding at the active site.

Lipases are able to hydrolyze ester bonds of insoluble acylglycerols and also to catalyze the inverse reaction, the synthesis of ester bonds from an alcohol moiety (e.g. provided by glycerol) and a carboxylic moiety (e.g. provided by a fatty acid). In addition, to hydrolysis and ester synthesis, lipases can also catalyze transesterification reactions (cleavage and formation of ester bonds in a sequential way). Reactions catalyzed by lipases are summarized in Figure 6.6. Some of the most studied hydrolysis reactions catalyzed by immobilized lipase in multiphase membrane reactors are summarized in Table 6.4.

Table 6.4 Hydrolysis reactions catalyzed by immobilized lipase in multiphase membrane reactors

Source of lipase	Method of immobilization and support	Membrane reactor (MR)	Substrate	Reference
Rhizopus nigricans	Cross-linking (PTFE)	Flat-sheet biphasic membrane	Sunflower oil olive oil	Rucka et al. (1991)
Candida rugosa	Gelified (Cuprophan)	Hollow-fibre biphasic MR	Triglycerides	Pronk et al. (1988)
Aspergillus niger	Adsorption (Polypropylene)	Flat-sheet biphasic MR	Glycerides of butter oil	Malcata et al. (1991)
Candida cylindracea	Adsorption (Polypropylene)	Plate-and-frame MR	Olive oil	Hoq et al. (1985)
Candida rugosa	Covalent (PVC, Chiton, Chitosan, Agarose)	Flat-sheet emulsion reactor	Plant oil	Shaw et al. (1990)
Rhizopus	Adsorption (PTFE)	Flat-sheet biphasic membrane	Plant oil	Rucka and Turkiewicz (1989)
Candida rugosa	Adsorption (DEAE-Sephadex)	Packed-bed organic monophase	Olive oil	Yang and Rhee (1992)
Rhizopus delemar		Lewis type cell[a]	2-Naphthyl ester	Hyake et al. (1991)

[a] Biphasic reactor without polymeric membrane.

6.2.2 *Membrane Modules Used in Biphasic Systems*

In this section we list and characterize modules used in biphasic systems.

Membrane modules of polyacrylonitrile (PAN)

These modules have been commercialized by Sepracor Inc. MA, USA. They are made of asymmetric capillary membranes with the following characteristics:

Molecular mass cut-off 50 kDa

Internal diameter of fibres $210(\pm1)$ μm

External diameter of fibres $310(\pm1)$ μm

Number of fibres in a module $8500(\pm1)$

Fibre length 133 mm

External membrane surface 0.7 m^2

Membrane modules of polyamide (PA)

Such modules have been realized in laboratories by assembling asymmetric capillary fibres in glass tubes. The membranes were provided by Forschungsinstitut Berghof, Germany. The characterstics of the modules are:

Molecular mass cut-off 10 or 50 kDa

Inner diameter 1($\pm$0.2) mm

External diameter 2($\pm$0.2) mm

Number of fibres 4

Fibre length 180 mm

External membrane surface area 45×10^{-4} m^2

Volume of membrane thickness 6.7 cm^3

Membrane modules of polypropylene (PP)

A microporous polypropylene membrane in hollow-fibre form is produced by Hoechst Celanese and sold under the trademark Celgard X10-400. The microporous structure is uniform throughout the inside of the membrane thickness; on the surface there is present a thin layer with slightly smaller pore size than inside. Further characteristics are:

Internal diameter 400 μm

Wall thickness 30 μm

Porosity 0.3

Pore dimensions 0.075×0.15 μm (width $\times$ length)

The same company also produces flat-sheet PP membranes (Celgard 2500) that can be assembled in flat-plate modules. For these:

Pore dimension 0.075×0.25 μm (width $\times$ length)

Other PP modules used as prepared in laboratories

Pore diameter 0.2 μm

Internal diameter 1.8($\pm$0.2) mm

External diameter 2.2($\pm$0.2) mm

Number of fibres 4

Fibre length 180 mm

Hollow fibre made of tetrafluoroethylene

PTFE hollow fibres are produced by Japan Gore-tex Co. Ltd. The characteristics of these membranes are:

Inner diameter 900($\pm$0.2) μm

Length of fibre 495($\pm$0.2) mm

Porosity 0.7

Specific activity ($10^3 \, \mu$mol $h^{-1} g^{-1}$)

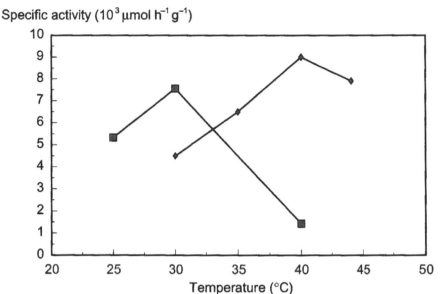

Figure 6.7   Effect of temperature on specific activity of immobilized lipase with different substrates: olive oil (♦) and an ester of arylpropionic acid (CNE) (■).

6.2.3   *Influence of Operating Conditions on the Performance of Biphasic Membrane Reactors*

The performance of biphasic membrane reactors is influenced by the operating conditions as well as by the properties of the immobilized enzyme and membrane properties. pH of the aqueous phase, ionic strength, temperature, concentration of immobilized enzyme, site of immobilization, concentration of substrate, viscosity of the organic phase, trans-membrane pressure, axial flow rate and the use of cocurrent or countercurrent flow regime exert strong influence on the overall efficiency of a biphasic membrane reactor. These parameters can affect the catalytic activity of the immobilized biocatalyst and mass transfer of reagents through the membrane system.

The optimum value of each parameter can not be known *a priori*, and if it is not already reported in the literature it must be determined experimentally. In general, optimal pH and temperature values do not change very much between free and immobilized lipase (unless enzyme is immobilized by chemical bonding). These values can change if the enzyme is used with different substrates or in different buffer solutions. For example, the optimal temperature value for free and immobilized lipase (from *C. rugosa*) is 40 °C when using olive oil as substrate, but is 28–30 °C when using an ester of arylpropionic acid (CNE) as substrate (Figure 6.7). The optimal pH and temperature values for several lipases are listed in Table 6.5.

Substrate viscosity influences the performance of the system by affecting mass transfer across the membrane up to the enzyme. Studies with low water solubility substrates of different viscosity have indicated that apparent reaction rate for enzyme-loaded systems is higher for substrates with lower viscosity. Experiments have been carried out using olive oil and CNE. Enzyme was used free in an organic/aqueous emulsion reactor and immobilized in an asymmetric capillary membrane (made of polyamide) and used in a

Table 6.5   pH and temperature optima for several lipases

Source of lipase	Status (immobilized or free)	Substrate	Aqueous phase	pH	$T(°C)$	Reference
Candida cylindracea or *rugosa*	Free and immobilized	Olive oil	Sodium phosphate 50 mmol l^{-1}	8.0	38–40	Giorno et al. (1995)
Candida cylindracea or *rugosa*	Free and immobilized	Ester of arylpropionic acid	Sodium phosphate 50 mmol l^{-1}	8.0	28	Giorno et al. (1997)
Candida cylindracea	Free	Olive oil		~5.0	37	Molinari et al. (1988)
Aspergillus niger	Immobilized	Butter oil		8.0	30	Malcata et al. (1991)
Candida rugosa	Immobilized	Oil		5.4	40	Shaw et al. (1990)
Candida rugosa	Immobilized	Soybean oil	Maleic acid	6.0	30	Pronk et al. (1988)
Candida rugosa	Free	Oil		7.5	35	Shaw et al. (1990)
Candida rugosa	Immobilized	Oil		8.5	45–55	Shaw et al. (1990)
Candida cylindracea	Immobilized	Soybean oil	Phosphate 0.1 mol l^{-1}	7.0	37	Tanigaki et al. (1993)

biphasic organic/aqueous system. Experiments were performed at pH 8.00 and a temperature of 40 °C. Figures 6.8 and 6.9 show performance for free and immobilized lipase, respectively, with both substrates.

Experimental results indicate that the decrease of reaction rate of immobilized enzyme with respect to the free enzyme is mostly due to the low mass transport of reagent across the enzyme-loaded membrane. Lipase free in emulsion showed higher reaction rate when using olive oil as substrate ($V_r = 6.954(\pm0.610)$ mmol dm^{-3} h^{-1}) than with CNE ($V_r = 1.123(\pm0.046)$ mmol dm^{-3} h^{-1}) (Figure 6.8). In contrast, lipase immobilized in the biphasic membrane reactor showed higher observed reaction rate when using CNE as substrate ($V_r = 0.129(\pm0.009)$ mmol dm^{-3} h^{-1}) than with olive oil ($V_r = 0.100(\pm0.008)$ mmol dm^{-3} h^{-1}) (Figure 6.9). In the immobilized system, the greater decrease of reaction rate with olive oil, a natural substrate for lipase, is due to the high viscosity of this substrate, which negatively affects the transport.

The flow rate of the organic phase (which contains the substrate) strongly influences the reactor performance. Figure 6.10 shows the observed reaction rate as a function of flow rate using olive oil as substrate. The reaction rate decreases with increase of the flow rate; similar behaviour is obtained using CNE as substrate, although the effect is less negative than with olive oil. At organic flow rate conditions of 80, 160 and 240 ml min^{-1} the resulting reaction rates were respectively, 0.100, 0.044 and 0.014 mmol dm^{-3} h^{-1} for

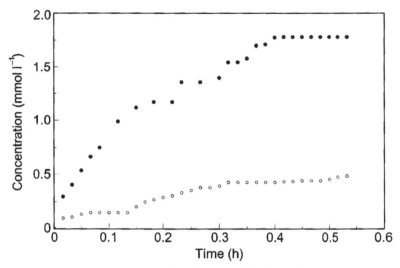

Figure 6.8　Volumetric reaction rate of free lipase with olive oil (•) and CNE (∘).

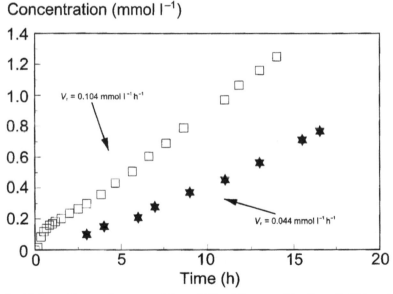

Figure 6.9　Volumetric reaction rate of immobilized lipase with olive oil (★) and CNE (□) as substrates.

olive oil and 0.129, 0.104 and 0.061 mmol dm^{-3} h^{-1} for CNE. The reactors were made of polyamide capillary membranes, with a membrane surface area of 45×10^{-2} m^2. The amount of immobilized lipase was about 6.7 mg, and transmembrane pressure was maintained at about 0.34 bar.

Increase of the flow rate of the aqueous phase (which extracts the product) positively influences the biphasic reactor performance. This is because the increase of axial flow rate of extractant phase increases mass transfer of product from the reaction microenvironment (enzymae–membrane) to the bulk. The removal of the product has the advantages of enhancing the reaction yield and reducing inhibition phenomena. The

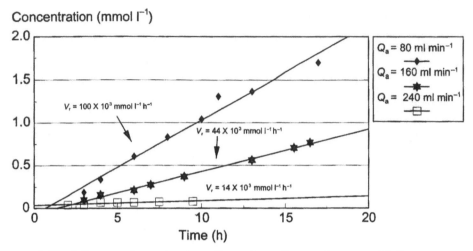

Figure 6.10 Influence of organic phase flow rate on the reaction rate with lipase immobilized in/on the sponge layer (immobilized protein = 6.7×10^{-3} g). Q_a = axial flow rate.

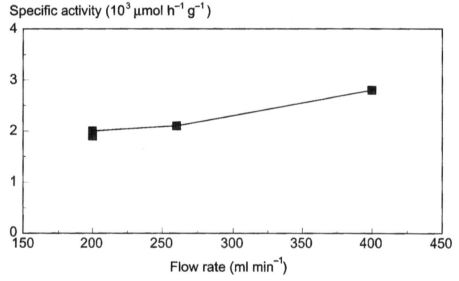

Figure 6.11 Influence of aqueous flow rate on the performance of biphasic membrane reactor with lipase immobilized in/on the sponge layer of asymmetric capillary membrane (immobilized protein = 12×10^{-3} g).

general behaviour of catalytic activity as a function of axial flow rate of the aqueous phase is shown in Figure 6.11.

In addition to the axial flow rate values of the organic and aqueous phases, the flow regime of both phases influences the biphasic membrane reactor's performance. The use of a countercurrent flow regime improves the reaction rate by increasing mass transfer efficiency. For example, when flowing the two phases in countercurrent regime along a hollow-fibre membrane reactor, an increase of 60% has been obtained over the cocurrent regime (Figure 6.12). On the other hand, in cocurrent flow the two phases are kept stable for longer periods of operation.

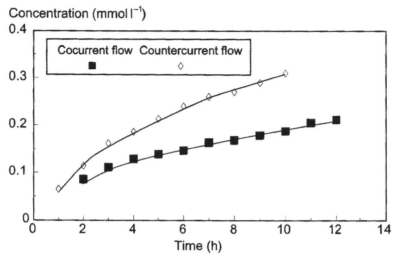

Figure 6.12 Fatty acid concentration as a function of time in biphasic reactor operated in cocurrent (■) or countercurrent flow (◊).

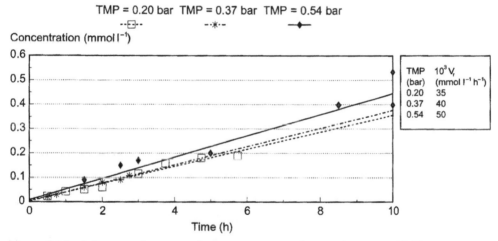

Figure 6.13 Influence of transmembrane pressure on the reaction rate with lipase immobilized in/on the sponge layer (immobilized protein = 3.3×10^{-3} g).

The influence of transmembrane pressure (positive on the membrane side containing the substrate) on the efficiency of biphasic membrane reactors also has to be considered. If working at values far below the breakthrough pressure, the effect of transmembrane pressure is not significant. On the other hand, there are ranges of pressure values that affect mass transport and thus reaction rate in two-separated-phase membrane reactors. When working in these ranges, an increase of transmembrane pressure yields an increase of reaction rate. This is particularly evident when enzyme is immobilized in the sponge layer of asymmetric membranes (Figure 6.13). There are two possible explanations for this behaviour: (1) increasing the oil phase pressure increases the penetration depth in the sponge layer of the membrane and more substrate comes into contact with the enzyme; (2) the reaction front is located deeper inside the sponge layer of the membrane, and the effective diffusion distance for the product to reach the aqueous phase decreases. Using

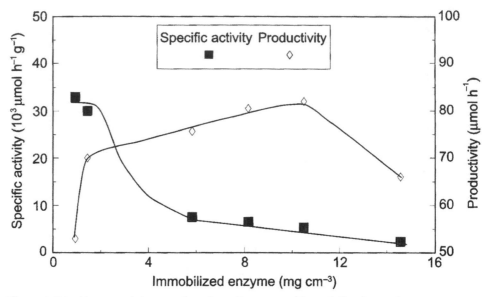

Figure 6.14 Lipase activity as a function of amount of immobilized protein.

lipase immobilized on the surface of the dense layer, a smaller increase of product concentration in the aqueous phase is generally observed compared with the case where the enzyme is immobilized in the sponge layer. This means that the latter configuration facilitates mass transfer from and to the reaction microenvironment. Furthermore, the higher rate with the enzyme immobilized within the sponge layer might be related to a better enzyme distribution; the enzyme may be more effective when present as an aqueous solution in the membrane pores than when gelified on the membrane surface.

The distribution of catalyst within the membrane support and the thickness of the protein layer formed strongly influence the transport of reagent and thus the reactor efficiency. The factors that determine the overall behaviour can be various. For example, the enzyme can distribute at different positions within the membrane, changing the distance through which the substrate must diffuse to reach the catalyst. During the immobilization procedure, proteins can be packed as layers, forming an additional barrier to transport and obstructing the catalyst on the rear layers. Furthermore, some enzymes (such as lipase) form inactive dimers when present at high concentration, reducing catalytic activity. On the basis of these considerations it can be assumed that specific activity ($mmol\ h^{-1}\ g^{-1}$) of immobilized enzymes remains constant until all the loaded protein is distributed as a monolayer — in other words, until all the catalyst can be reached by the substrate — whilst it decreases as the previously described factors become limiting. The optimal value of immobilized mass per unit of membrane volume depends on enzyme characteristics, immobilization procedure, membrane structure and porosity, transport mechanism, and so on. In Figure 6.14 the optimal range of amount of lipase immobilized in a biphasic hollow-fibre membrane reactor is reported. The membrane was made of polyamide, with asymmetric structure; the enzyme was immobilized in the sponge layer by crossflow ultrafiltration. The range of enzyme concentration values in which the specific activity remains constant must be known, for example, for measuring kinetic parameters. In practice, in this range the system works under reaction-limited conditions, rather than diffusion-limited conditions. In these cases, the kinetics of immobilized systems can be interpreted using a Michaelis–Menten model.

Table 6.6 Effect on reaction rate of toluene entrapped in a PAN membrane

Crude protein (mg)	Protein concentration by BCA (mg)	Immobilized protein (mg)	Retention (%)	$\mu mol\ h^{-1}$	$\mu mol\ h^{-1}\ mg^{-1}$ (crude)	$\mu mol\ h^{-1}\ mg^{-1}$ (BCA)
*	1730	1227	70	5821	n.a.	4.76
* 4750	970	467	48	n.a.	n.a.	n.a.
* 4750	1017	438	43	5120	1.13	10.28
# 4750	1235	503	41	5715	1.2	11.36
# 4500	1232	665	54	9296	1.95	13.98
# 4500	1121	658	58	9600	2.1	14.58

*, experiments without toluene entrapped in the membrane.
#, experiments with toluene entrapped in the membrane.
Retention (%) = mg immobilized protein/mg initial protein (BCA) × 100.
n.a. = not available.

6.2.4 Enhancement of Transport Rate across Enzymatic Membranes in Biphasic Reactors

Mass transfer across the enzyme-loaded membrane is the factor that most influences the efficiency of biphasic membrane reactors. To improve the productivity of these systems, mass transfer of reagents must be improved. Besides operating at the optimal values of parameters previously discussed, mass transfer can be improved by modifying the environment through which the reagents must diffuse. One way to do this it is to modify the liquid phase contained within the membrane, to realize a diffusion path with high affinity for the reagents. For example, in a biphasic organic/aqueous membrane reactor with hydrophilic enzymatic membranes that separate the two phases, the low water solubility substrate must diffuse from the organic phase, through the hydrophilic membrane (containing water as inner liquid phase) and reach the enzyme. If the inner liquid phase of the membrane is constituted by both phases, the diffusion of the low water solubility substrate will be improved by the presence of the organic phase. The drawback of such a system is that the membrane's capacity to keep the two phases separated is reduced, since the presence of organic phase reduces the breakthrough pressure limit. Nevertheless, when enzymes are loaded onto membranes having organic/aqueous inner liquid phase, the separation properties of the membrane can be restored.

The properties of a hydrophilic membrane with an inner liquid phase comprising organic:water 50:50 will be discussed as an example of this kind of system (Giorno and Lopez, 1992). A Sepracor biphasic organic/aqueous membrane system (MBR 500) was used to carry out the experiments. It includes a polyacrylonitrile membrane bioreactor and two panels to control flow conditions. Amyl acetate, used as substrate, was recirculated along the shell side; the acid produced was recovered in the buffer solution recirculated along the lumen side, and measured by titration. Capillary membrane modules made of polyacrylonitrile were treated to produce a membrane with a mixture of toluene:water 50:50 as the inner liquid phase. Lipase was immobilized on such treated membranes and the reaction rate was measured. A comparison between the performances of membrane bioreactors containing or not containing toluene in the inner liquid phase is reported in Table 6.6. When the toluene was present in the membrane, the specific activity ($\mu mol\ h^{-1}\ mg^{-1}$ protein measured by the BCA test from Pierce) was increased by

Table 6.7 Effect on the transport rate of toluene entrapped in a PAN membrane

System	Immobilized protein (mg)	Transport rate (μmol min^{-1})
PAN membrane		28.80
PAN membrane + enzyme	439	15.28
PAN membrane + enzyme + toluene	521	56.84

42% with respect to the case where the toluene was not present. The effect of toluene entrapped in the inner membrane liquid phase on the transport rate in the absence of enzymatic reaction was also investigated.

The transport rate of ibuprofen from the organic phase (toluene, recirculated along the shell) into the aqueous phase (phosphate buffer or distilled water, recirculated along the lumen) was used as a model system. When the toluene was not present in the membrane, there was no difficulty keeping the two phases separated, since the hydrophilic membrane completely separated the organic phase from the aqueous phase. When the toluene was present inside the membrane, the toluene, recirculated in the shell side, ultrafiltered across the membrane. The permeate flow rate of the toluene was the same as that of the water (30 ml min^{-1}), which was evidence that the organic phase as well as the aqueous phase was entrapped in the membrane. When the enzyme was also present in the membrane, the two phases remained completely separated, even if the toluene was present inside the membrane.

The transport rate of the ibuprofen across the membrane was then measured using (a) the polyacrylonitrile membrane itself; (b) the polyacrylonitrile membrane loaded with enzyme; and (c) the polyacrylonitrile membrane with toluene and enzyme loaded. As shown in Table 6.7, the transport rate is higher when the toluene is present in the membrane, even compared with the membrane without enzyme loaded.

6.2.5 *Biphasic Membrane Reactors in Oil Treatment: Case Study*

Recently, the technology of oils and fats has been studied to assess the possibility of replacing traditional chemical processes with biotechnological ones. Enzyme membrane reactors, which can contribute to the solution of problems such as stability of catalyst, product inhibition phenomena, and selective removal of products, can be applied in oil treatment. The use of lipase immobilized in membrane reactors for oil treatment is well documented in the literature; some examples are illustrated in Table 6.8.

Depending on the physicochemical properties of reaction mixtures, different reactor configurations may have different performance, which cannot easily be predicted and has to be studied case by case. A comparison of performance of lipase from *C. rugosa* used in different reactor configurations is reported below.

Reactor configurations

Stirred tank reactor (STR) The reaction mixture was formed of 18 ml of phosphate buffer at pH 8, 1 ml of substrate, and 2 ml of enzyme solution 500 μg ml^{-1}. The total reaction volume was 21 ml in a batch reactor of 50 ml. The enzyme concentration in the reaction mixture was 4.7×10^{-3} % (w/v).

Table 6.8 Applications of biphasic enzyme membrane reactors in oil treatment

Enzyme	Membrane reactor type	Application
Lipase (*C. cylindracea*)	Hydrophobic plate-and-frame	Hydrolysis of triglycerides of olive oil (Hoq and Yamane, 1985)
Lipase (*C. cylindracea*)	Hydrophilic hollow-fibre	Hydrolysis of soybean oil (Pronk et al., 1988)
Lipase (*A. niger*)	Hydrophobic flat-sheet	Hydrolysis of glycerides of butter oil (Malcata et al., 1991)
Lipase (*R. nigricans*)	Hydrophilic and hydrophobic	Hydrolysis of plant oils (Rucka et al., 1991)
Lipase (*A. niger*)	Hydrophobic hollow-fibre	Hydrolysis of glycerides of butter oil (Malcata et al., 1992a)
Lipase (*C. cylindracea*)	Hydrophobic hollow-fibre	Hydrolysis of triolein (Goto et al., 1992)
Lipase (*C. cylindracea*)	Hydrophilic/hydrophobic flat-sheet	Hydrolysis of soybean oil (Tanigaki et al., 1993)
Lipase (*C. cylindracea*)	Hydrophilic hollow-fibre	Hydrolysis of triglycerides of olive oil (Molinari et al., 1994)
Lipase (*C. cylindracea*)	Hydrophilic hollow-fibre	Hydrolysis of triglycerides of olive oil; enantioselective hydrolysis of (*R,S*)-esters (Giorno et al., 1995)

Emulsion enzyme membrane reactor (E-EMR) (Figure 6.15a) The total volume of reaction mixture was 500 ml, containing 3% (v/v) of olive oil, and 4×10^{-3} % (w/v) of enzyme. The reaction mixture was emulsified in the tank reservoir and supplied by a peristaltic pump to the ultrafiltration membrane module. The module was made of polyamide capillary membrane with asymmetric structure, a nominal molecular mass cut-off (NMWCO) of 10 kDa, inner/outer diameter of 1.5/2.5 mm, and a total inner membrane surface area of 25.4 cm². The membranes were kindly supplied by Berghof, Germany. The emulsion was recirculated along the lumen circuit (inner path of fibres) at a flow rate of 1.6 ml min⁻¹ and transmembrane pressure (TMP) of 0.6 bar. In this reactor the enzyme was compartmentalized by the membrane in the emulsion, while the water phase ultra-filtered through the membrane module. The volume in the tank was kept constant by adding fresh water phase.

Biphasic organic/aqueous enzyme membrane reactor (B-EMR) (Figure 6.15b) Polyamide capillary membranes with NMWCO of 50 kDa, inner/outer diameter 1/2 mm, and a total external membrane surface area of 45 cm² were used. The two immiscible phases were separately recirculated along the two different circuits of the membrane module: the organic phase was recirculated (at 1.6, 2.5 and 150 ml min⁻¹) along the side containing the enzyme. The aqueous phase was phosphate buffer at pH 5 or 8, as indicated in the experimental section; it was recirculated at a flow rate of 167 and 350 ml min⁻¹. The reaction and separation of products occurred at the enzyme-loaded membrane interface, where the two phases come into contact.

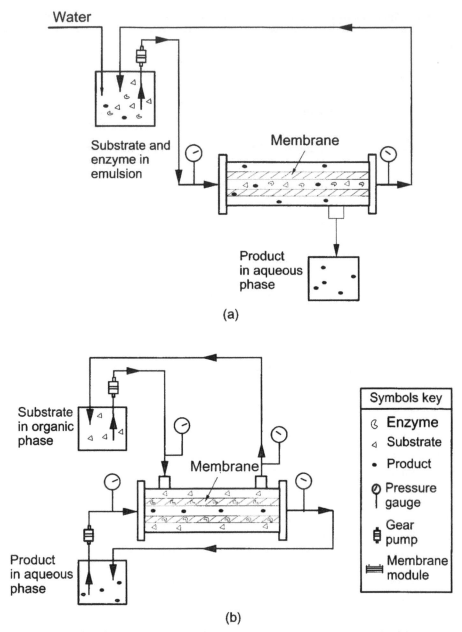

Figure 6.15 (a) Emulsion-enzyme membrane reactor (E-EMR) (separation by UF).
(b) Biphasic enzyme membrane reactor (B-EMR) (separation by membrane-based solvent extraction).

The enzyme was loaded on/in the membrane by physical or chemical methods. In the first case, the enzyme was immobilized by crossflow filtration from the shell to the lumen; in this way, the enzyme remained entrapped in the sponge layer of the asymmetric membrane, its molecular mass (67–70 kDa) being higher than the NMWCO (50 kDa) of the membrane. The amount of protein immobilized was calculated by mass balance between the initial and final mass (mass in the permeate + mass in the retentate).

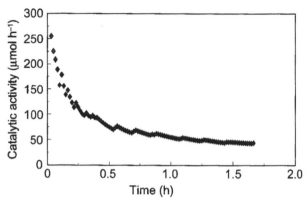

Figure 6.16 Catalytic activity of free lipase in STR system.

The procedure used to covalently bind the enzyme to the membrane by means of the glutaraldehyde was as follows: 2.5% w/v glutaraldehyde solution in 500 ml of carbonate buffer 0.1 mol l^{-1}, pH 10, is recirculated along the shell side for 3–4 h at 34 °C at an inlet pressure of 0.1 bar. After washing the unreacted glutaraldehyde with water, the enzyme solution is recirculated for about 15 h at an inlet pressure of 0.1 bar and a temperature between 5 and 10 °C to reduce enzyme deactivation. After washing the non-bound enzyme with water, the enzyme membrane module is ready to be tested.

Using the optimal reaction conditions for lipase (pH 8.00 and $T = 40$ °C) determined in previous work (Giorno and Drioli, 1997; Giorno et al., 1997), the catalytic activity of lipase in an emulsion stirred tank reactor was measured (Figure 6.16). The amount of enzyme used was 5×10^{-3} % (w/v) (1 mg in 20 ml).

Owing to mechanical stirring and/or to product inhibition, the catalytic activity of free lipase in the STR decreased by 50% in about 15 min. This inhibition proved to be irreversible: the equilibrium was not shifted by adding fresh substrate.

Lipase has also been studied suspended free in an E-EMR, where the membrane compartmentalizes the enzyme in the reaction vessel while the water phase (and the water-soluble products) is removed by ultrafiltration. As the water phase permeates through the membrane, fresh water phase is added to the emulsion. In this way, the water-soluble product concentration in the reaction mixture is decreased by diafiltration, reducing product inhibition.

During the reaction, the water phase (Tris-HCl 0.1 mol l^{-1}, pH 6 buffer solution) containing 4×10^{-3} % (w/v) of enzyme (20 mg in 500 ml) was recirculated along the external surface of the polyamide capillary membrane (cut-off 10 kDa) for about 15 min. After this time the substrate was added and the reaction started. The ultrafiltration was carried out from shell to lumen using a TMP of 60 kPa. The emulsion was recirculated at a flow rate of 1.6 ml min^{-1}.

Figure 6.17 shows the degree of conversion $[X_t = (C_{t0} - C_t/C_{t0})]$ versus time. The initial reaction rate was about 6.8 mmol l^{-1} h^{-1} (Molinari et al., 1994). As observed for the STR, the enzyme activity decreased. However, in this case, inhibition is reversible: the equilibrium is shifted to the right by adding fresh substrate.

Using this kind of configuration, the separation of glycerol into the aqueous phase (pH 6) is obtained by diafiltration at constant volume, with the drawback that the product is present in a very diluted form and needs subsequent concentration. Furthermore, it has been calculated that during diafiltration most of the enzyme is gelified on the membrane.

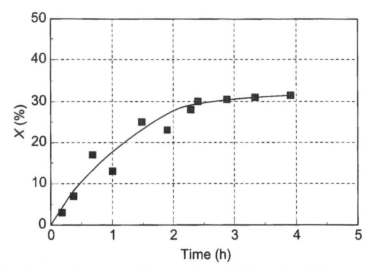

Figure 6.17   Degree of conversion (X) versus time with lipase in E-EMR.

To overcome problems encountered with the previous two reactor configurations, a biphasic organic/aqueous enzyme membrane reactor (B-EMR) has been investigated. In this reactor, the two liquid phases are kept separated and in contact with the membrane, which also contains the enzyme entrapped in the pores or gelified on the surface. The two phases are recirculated along the two sides of the membrane, with the organic phase always on the side containing the catalyst.

Because hydrophilic membranes have been used, a positive transmembrane pressure (below the breakthrough pressure limit) from organic to aqueous phase is necessary in order to avoid leakage of water into oil. The reaction and separation of product occur at the interface where the two immiscible phases come into contact.

The B-EMR has been investigated with the enzyme immobilized by physical or chemical methods. In the first case, the enzyme was physically entrapped in the sponge layer of a capillary membrane (PA with cut-off = 50 kDa) by crossflow ultrafiltration. Most of these experiments have been carried out using 150 ml of lipase solution (50 µg ml^{-1} in 0.1 mol l^{-1} sodium phosphate buffer, pH 5). The amount of enzyme immobilized was about 2.3 mg. The experiments were carried out using cocurrent flow regime. The aqueous phase (200 ml of phosphate buffer, pH 5) was recirculated at a flow rate of 167 ml min^{-1} in the lumen side of the fibres and the organic phase (225 ml of olive oil) in the shell at a flow rate of 1.6 ml min^{-1}. At pH 5 only the glycerol is present in the water phase, the fatty acids being more soluble in the oil phase. In this way, the hydrolysis and the separation of fatty acids from the glycerol is obtained simultaneously. Under these conditions the apparent volumetric reaction rate was about 4.5 mmol l^{-1} h^{-1}.

Using the same configuration, experiments have been carried out using a phosphate buffer solution at pH 8 as aqueous phase. At this pH value the fatty acids are also soluble in the aqueous phase. The amount of enzyme loaded into the reactor was about 3 mg. The experiments were performed recirculating 600 ml of phosphate buffer along the lumen at a flow rate of 350 ml min^{-1}, and 200 ml of olive oil along the shell at a flow rate of 150 ml min^{-1}. The apparent volumetric reaction rate was calculated by the product concentration present only in the aqueous phase, and was about 0.17 mmol l^{-1} h^{-1}. The lower reaction rate obtained in this last case is probably due to the higher organic phase flow

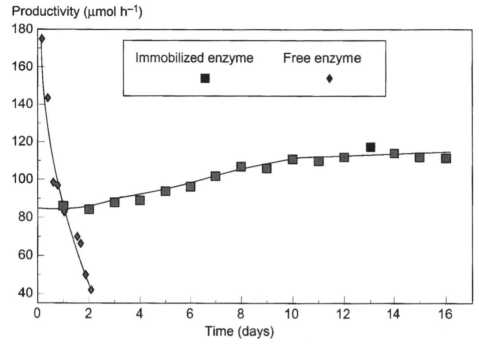

Figure 6.18   Productivity vs time of free (♦) and immobilized lipase (■) using olive oil as substrate.

rate, which negatively affects the reactor performance. Moreover, it is possible that part of the enzyme is extracted into the organic phase at higher crossflow velocity, with an important decrease of enzyme activity.

In another series of experiments the lipase was chemically immobilized on the sponge side of capillary fibres (with 50 and 10 kDa cut-off) by pretreatment with glutaraldehyde. The amount of enzyme immobilized on 50 and 10 kDa membranes was about 1.9 mg and 1.3 mg, respectively. Experiments were carried out by recycling the oil phase at 1.35 (or 2.15) ml min^{-1} and the aqueous phase at 167 (or 500) ml min^{-1}. The reaction rate was about 4.5 mmol l^{-1} h^{-1} with the 50 kDa membrane and about 3 mmol l^{-1} h^{-1} with the 10 kDa membrane. This decrease is probably due to the higher barrier to the mass transport caused by the low pore dimensions of the 10 kDa membrane.

In all the different conditions, the catalytic stability of the immobilized enzyme was much higher than that of free enzyme. The two systems are compared in Figure 6.18 (Giorno et al., 1995). The experiments were carried out using phosphate buffer at pH 8 as aqueous phase, the other reaction conditions were as indicated in the figure.

The use of enzyme membrane reactors for the hydrolysis of triglycerides present in olive oil and simultaneous separation of reaction products has been discussed. The reaction has been carried out using lipase free in suspension in a stirred tank reactor, in an emulsion enzyme membrane reactor, or immobilized in a biphasic enzyme membrane reactor. Under optimum operating conditions, the experimental results showed that the apparent volumetric reaction rate of the free enzyme (6.8 mmol l^{-1} h^{-1}) is higher than that of the immobilized enzyme (4.5 mmol l^{-1} h^{-1}). On the other hand, the catalytic stability is much higher for the enzyme immobilized in the B-EMR. Taking into the account that the separation of the product is simultaneously obtained as the reaction proceeds, it seems more convenient to use the biphasic enzyme membrane reactor. In this system the reaction rate is limited by the low mass transport of reagents through the enzymatic membrane.

6.3 Enzyme Membrane Reactors in Pure Organic Phase

As enzymes are catalysts of biological origin, they have traditionally been used in aqueous phases. However, in addition to the use of enzymes in biphasic organic/aqueous systems, through the 1990s the use of enzymes in pure organic solvents has increased considerably. This use provides new enzyme applications and methods for better understanding of the fundamentals of enzyme mechanisms. Enzymes in organic media become rigid and this rigidity enables the study of the properties of their active sites. The activity of enzymes in organic/aqueous systems is not surprising, since the biocatalyst is still present in the aqueous phase and the organic phase serves only to supply the substrate. When enzymes are used in organic solvents, they are able to work in microenvironments that contain only a few per cent of water (usually less than the solubility limit). Although this is surprising, numerous studies have confirmed that it is possible to carry out biotransformations in organic media. Klibanov and Zaks (1985), Zaks and Klibanov (1984, 1988), and Klibanov (1990) studied enzymatic catalysis in such non-natural media. They demonstrated that enzymes in organic solvents exhibit novel properties, such as enhanced stability and altered substrate specificity. For example, the α-chymotrypsin enzyme in aqueous media catalyzes the hydrolysis of peptides, while in anhydrous organic solvents (with a water content less than 0.01%) it catalyzes the transterification of amino acid esters.

Reactions catalyzed by lipase in organic media are summarized in Table 6.9.

In general, hydrophobic rather than water-miscible organic solvents are more suitable reaction media. This is because the hydrophobic solvent does not extract the water molecules necessary to the enzyme catalytic activity. Study of the influence of solvent hydrophobicity on the water content of enzymes, such as α-chymotrypsin, lipases and oxidoreductases, showed that the less hydrophobic the solvent and the lower the water content of enzymes, the lower the catalytic activity. This means that the interactions between the solvent and the amount of water surrounding the enzyme influence the catalytic behaviour more than those between the solvent and the enzyme itself.

Table 6.9 Lipase-catalyzed reactions in organic media

Lipase	Medium	Reaction	Reference
	Apolar solvents	Peptide synthesis	West and Wang (1987)
Porcine pancreas	Chloroform	Glycerol esterification	Martins et al. (1993)
Candida antartica	*In vacuo* (immobilized by Novo-Nordisk)	Esterification of glycerol and interesterification of tributyrin with EPA[a] and DHA[a]	Haraldsson et al. (1993)
Pseudomonas cepacea	Tetrahydrofuran	Flavone alcoholysis	Natoli et al. (1990)

[a] EPA = eicosapentaenoic acid; DHA = docosahexaenoic acid.

The use of enzymes in organic solvents is interesting owing to their enhanced thermostability (Zaks and Klibanov, 1984), conformational rigidity (Zaks and Klibanov, 1988), and novel properties such as specificity for substrates different from the natural ones. Enzymes in organic solvents are used to catalyze reactions that are not feasible in aqueous systems owing, for example, to unfavourable thermodynamic equilibria or low solubility of reagents.

Hydrolases, and in particular lipases, are used in organic solvents to catalyze a variety of reactions — esterification, alcoholysis, aminolysis — that in aqueous systems are prevented by hydrolysis. For example, the mechanism of interesterification reactions, at molecular level, involves hydrolysis of the ester molecule followed by an esterification reaction. As a result, water must be present in catalytic amounts; in fact, since the processes in question are reversible, the presence of very large amounts of water promotes the hydrolysis reaction.

In general, the advantages of using enzymes in organic phases can be summarized as follow:

- High solubility of nonpolar substrates not soluble in water;
- Realization of processes which are thermodynamically unfavourable in water (ester synthesis, etc.)
- High reagent concentration
- High product concentration
- Reduction of product and substrate inhibition
- Easy separation of products
- Enhanced thermal stability
- No microbial contamination

On the other side, the main disavantage is the deactivation of enzyme in organic solvents, especially in polar organic solvents.

6.3.1 *Immobilized Enzymes in Organic Solvents*

As already pointed out, the stability of enzymes in organic media depends strongly on the hydrophobicity of the solvent. The higher the polarity, the lower the stability. The polarity of various organic solvents is reported in Table 6.10. Nevertheless, the stability of enzymes in organic media can be enhanced by using immobilized enzymes. Among other enzymes, lipase has been the most studied. Lipase from *Candida rugosa* has been immobilized on DLHE-Sephadex and Amberlite.

The stability of enzymes used in organic media can be enhanced by using stabilizing compounds, modifying enzyme molecules and immobilizing enzyme in/on insoluble supports.

Lipase–surfactant complexes have been reported by Basheer et al. (1995a, 1995b) and Isono et al. (1995a, 1995b). These complexes have been prepared by mixing aqueous solutions of lipase (from *Pseudomonas* sp., *Rhizopus japonens*) and ethanol solutions of sorbitan monostearate. The lipase showed good solubility, catalytic activity and thermal stability in organic media such as n-hexane. The modified enzyme was also used in a stirred tank reactor with polyimide flat-sheet membrane (NITTO) for reported batch

Table 6.10 Characteristics of commonly used organic solvents

Solvent	$\log P^a$
Polar	
Dioxane	−1.1
N,N-Dimethylformamide	−1.0
Methanol	−0.76
Acetonitrile	−0.33
Ethanol	−0.24
Acetone	−0.23
Tetrahydrofuran	0.49
Dichloromethane	0.60
Ethyl acetate	0.68
2-Pentanone	0.80
Butanol	0.80
Ethyl ether	0.85
1,2-Dichloroethane	1.2
Butyl acetate	1.7
Isopropyl ether	1.9
Chloroform	2.0
Benzene	2.0
Apolar	
Toluene	2.5
Butyl ether	2.9
Hexane	3.5
Isoctane	4.5
Decane	5.6
Dodecane	6.6

[a] $\log P$ is the logarithm of the partition coefficient of a given compound in the octanol–water two-phase system.

synthesis of wax ester. Higher conversions were mantained during repeated experiments (Isono et al., 1995a, 1995b).

Lipases from various microorganisms have been chemically modified with poly(ethylene glycol) derivatives by Kodera et al. (1994). Since the polymers are amphipathic, the resulting modified lipases were soluble in hydrophobic solvents. They were also thermostable, highly active, and preferentially catalyzed the esterification of (*R*)-isomers of secondary alcohols.

An organosoluble biocatalyst consisting of a noncovalent complex of an enzyme with a sugar-based amphiphilic polymer has been reported by Blinkovsky et al. (1994). The favourable microenvironment created around the surface of the enzyme molecule by hydrophilic carbohydrate moieties of the polymer increased the activity of subtilisin solubilized in organic solvents.

The modified enzyme shows enhanced activity and stability in organic solvents; nevertheless, when using polar solvents, the stability of the enzyme decreases. For this kind of solvent the use of immobilized enzyme is recommended. Enzymes have been immobilized on solid supports as illustrated in Table 6.11. It has been shown that not only is the stability improved, but also that a change in selectivity can be induced by immobilization (Jansen et al., 1990).

Table 6.11 Immobilized lipases used in organic solvents

Lipase source	Immobilization	Support	Substrate	Organic solvent	Reference
Candida cylindracea	Adsorption	DEAE Sephadex A50 Sephadex 450	Olive oil	Isoocatane	Yang and Rhee (1992)
Rhizomucor miehei	Adsorption	Controlled-pore glasses	Oleic acid Octan-1-ol		Bosley and Clayton (1994)
	Covalent bonds	Chitosan, PVC			Shaw et al. (1990)
Candida rugosa	Adsorption	Sephadex			Kang and Rhee (1989)
Pseudomonas cepacea	Adsorption	Zirconia	Flavone	Tetrahydrofuran	Natoli (1994)
Porcine pancreas	Adsorption	Celite	Dodecanol, decanoic acid	Hydrocarbons, ethers, ketones, etc.	Valivety et al. (1991)

Most studies on immobilized enzymes in organic phase have been carried out with biocatalysts immobilized in/on carrier dispersed in the reaction media as in STR systems or as used in packed-bed reactors (Malcata et al., 1992a, 1992b; Mushanta et al., 1993; Reyes and Hill, 1994; Bosley and Clayton, 1994; Shaw et al., 1990).

When the reaction medium consists of a polar organic solvent, the use of a hydrophilic matrix that keeps water around the enzyme will help to protect the biocatalyst. Furthermore, a high repartition coefficient will be obtained with increased catalytic activity (Yang and Rhee, 1992).

The use of immobilized enzyme membrane reactors with pure organic solvents has not yet been exploited practically. As far as the authors know, only a few studies have been carried out (Natoli, 1994).

Inorganic membranes have been used because they are more resistant to all organic solvents compared to organic membranes, which are stable only with few solvents (see Table 6.1.

Lipase from *Pseudomonas cepacea* in tetrahydrofuran has been used as a model system to study the catalytic behaviour of the immobilized enzyme in terms of activity, amount of loaded enzyme, and regioselectivity. *Pseudomonas cepacea* lipase of commercial grade (from Amano Enzyme Europe Ltd) was used. Its specific activity was 30 Units mg^{-1}, proteic content 2.6% (w/w) and molecular mass 33 kDa. Isoelectric points were pH 6.2 and 5.7, optimal pH was 7, and optimal temperature 50 °C.

Tubular zirconia ultrafiltration membranes with the following characteristics were used:

Support inner diameter 6 mm

External diameter 10 mm

Effective membrane length 190 mm

Membrane surface area 60 cm^2

Volume of membrane wall 10 cm^3

Mean pore size of carbon support 0.2 μm

On the inside of the tubular membrane, a selective layer of zirconium oxide was cast inside the macropores of the polycrystalline carbon tube (which forms the support material of the membrane). The selective layer had the following properties:

Specific area 66 m^2 g^{-1}

Density 5.6 g cm^{-3}

Mean particle diameter 16 nm

Molecular mass cut-off 10 kDa.

The membrane was mounted in a Carbosep module from Tech-Sep. Lipase was immobilized in the wall of the tubular inorganic membrane and the organic solution was continuously ultrafiltered through the membrane in a crossflow filtration mode. The system operated in a batch recycle mode, which means that the permeate was recycled back to the retentate stream.

The substrate of the regioselective enzymatic transesterification was 5,7-diacetoxyflavone to produce a partially acetylated flavone. Chemical synthetic methods for the acetylation of polyhydric molecules produce the totally acetylated form without regioselective protection of hydroxyl groups. The enzyme-catalyzed transesterification procedure for the protection–deprotection of the hydroxyls of the flavones using vinyl acetate as acetylating reagent gave no reaction, as reported by Natoli et al. (1990). The alternative route of the selective deacylation of one ester group in a flavone paracetate using butanol as transesterifying agent allowed a regioselective reaction to be obtained.

A scheme of this reaction is reported in Figure 6.19. The intermediate product (5-acetoxy-7-hydroxyflavone) is the new substrate for the subsequent reaction giving the product of total alcoholysis (5,7-dihydroxyflavone). Lipase free in suspension is able to catalyze both reactions, while immobilized lipase was able to catalyze only the first reaction step, producing only the product of partial alcoholysis, the 5-acetoxy-7-hydroxyflavone, a compound that does not exist in nature. With immobilization, the lipase lost the selectivity for one of the substrates; it was then possible to achieve a high regioselectivity towards the other substrate.

In the study of Natoli (1994) the behaviour of the alcoholysis reaction rate for suspended (Figure 6.20) and immobilized (Figure 6.21) lipase in anhydrous tetrahydrofuran was reported. As shown in the figures, the reaction rate as a function of amount of enzyme differs for the suspended and immobilized forms. For the free form the reaction rate increases as the concentration of enzyme increases, whilst for the immobilized form the reaction rate decreases as the enzyme concentration increases. These results are an effect of mass transfer properties of the different systems. In the immobilized system, the concentration of enzyme is a key factor affecting the performance because above a certain limit it can work as a barrier to the transport of reagents to the catalyst molecules, decreasing the observed reaction rate and catalytic activity. In the reaction system with the enzyme freely dispersed in the reaction medium, increase of enzyme concentration allows an increase of reaction rate simply because more catalyst is available to the substrate.

The highest reaction rate within the range of concentrations studied was found using immobilized enzyme (1.5 g cm^{-3}). The catalytic activity and stability of immobilized and

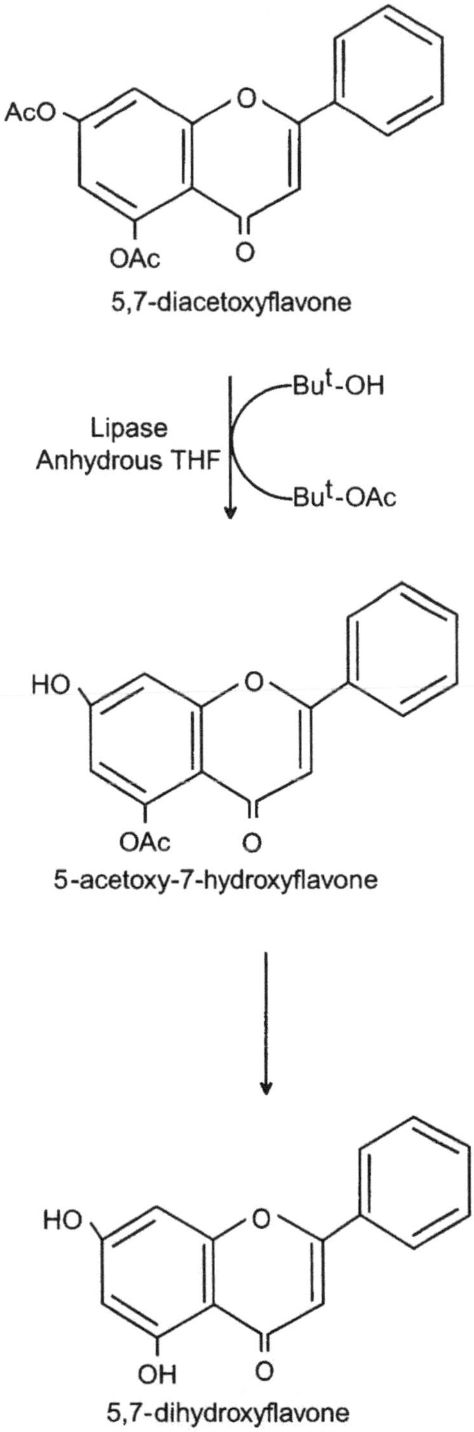

Figure 6.19 Alcoholysis reaction in organic media.

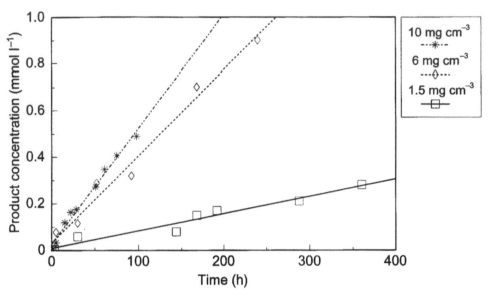

Figure 6.20 Reaction product as a function of time using different amounts of suspended (free) lipase in pure organic phase.

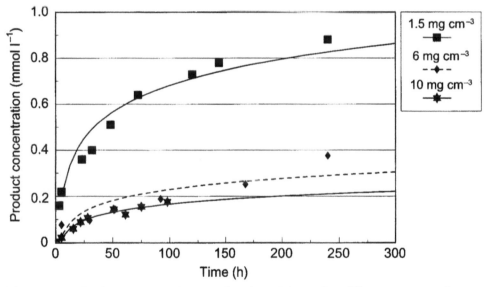

Figure 6.21 Product concentration as a function of time using different amounts of immobilized lipase in pure organic phase membrane reactor.

suspended enzymes were also compared (Figure 6.22). It can be seen that immobilized enzyme is more active than free enzyme and, after an initial decrease, the activity remains constant. This behaviour can be explained by the fact that the membrane prevents denaturation of the enzyme by forcing macromolecules into a physical microenvironment in which the tridimensional structure cannot unfold. The same phenomenon is probably responsible for the loss of selectivity for the intermediate substrate or, the other way around, for the improvement of regioselectivity for the initial substrate.

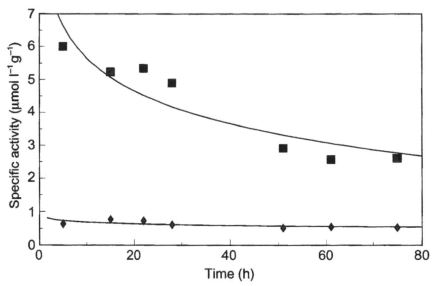

Figure 6.22 Specific activity as a function of time for immobilized (■) and suspended (free) (♦) lipase in pure organic phase (lipase concentration = 10 mg cm⁻³).

6.4 References

ADAMS, D.M. and BRAWLEY, T.G., 1981, *J. Food Sci.*, **46**, 677–680.

BASHEER, S., MOGI, K.-I. and NAKAJIMA, M., 1995a, Interesterification kinetic of triglycerides and fatty acids with modified lipase in n-hexane, *J. Am. Oil Chem. Soc.*, **72**(5), 511–518.

BASHEER, S., MOGI, K-I. and NAKAJIMA, M., 1995b, Surfactant-modified lipase for the catalysis of the interesterification of triglycerides and fatty acids, *Biotechnol. Bioeng.*, **45**, 187–195.

BENZONANA, G. and ESPOSITO, S., 1971, *Biochim. Biophys. Acta*, **231**, 15–22.

BLINKOVSKY, A.M., KHMELMITSKY, Y.L. and DORDICK, J.S., 1994, Organo-solvent biocatalyst for non-aqueous media, *Biotechnol. Tech.*, **8**(1), 33–38.

BOSLEY, S.H. and CLAYTON, J.C., 1994, Blueprint for a lipase support: use of hydrophobic controlled-pore glasses as model systems, *Biotechnol. Bioeng.*, **43**, 934–938.

CANIONI, P., JULIEN, R., RATHELOT, J. and SARDA, L., 1978, *Lipids*, **12**, 393–397.

CHOPINEAU, J., MELAFFERTY, F.D., THERISOD, M. and KLIBANOV, A.M., 1988, *Biotechnol. Bioeng.*, **31**, 208–214.

DESNELLE, P., 1972, in *The Enzymes*, Vol. 7, pp. 599–601, Academic Press, New York.

GARCIA, H.S., HILL, C.G. and AMUDSON, C.H., 1991, *J. Food Sci.*, **56**, 1233–1237.

GIORNO, L. and DRIOLI, E., 1997, Catalytic behaviour of lipase free and immobilized in biphasic membrane reactors with different low water-soluble substrates, *J. Chem. Tech. Biotechnol.*, **69**, 11–14.

GIORNO, L. and LOPEZ, J.L., 1992, Enhancement of transport rate by entrapment of water immiscible organic solvents in a hydrophilic membrane, Internal report, Sepracor Inc., USA.

GIORNO, L., MOLINARI, R., DRIOLI, E., BIANCHI, D. and CESTI, P., 1995, Performance of a biphasic organic/aqueous hollow fibre reactor using immobilized lipase, *J. Chem. Tech. Biotechnol.*, **64**, 345–352.

GIORNO, L., MOLINARI, R., NATOLI, M. and DRIOLI, E., 1997, Hydrolysis and regioselective transesterification catalyzed by immobilized lipases in membrane bioreactors, *J. Membr. Sci.*, **125**, 177–187.

GOTO, M., GOTO, M., NAKASHIO, F., YOSHIZUKA, K. and INOUE, K., 1992, *J. Membr. Sci.*, **74**, 207–214.

HARALDSSON, G.G., GUDMUNDSSON, B.O. and ALMARSSON, O., 1993, The preparation of homogeneous triglycerides of eicosapentaenoic acid and docosahexaenoic acid by lipase, *Tetrahedron Lett.*, **34**(36), 5791–5794.

HOQ, M.M. and YAMANE, T., 1985, Continuous hydrolysis of olive oil by lipase in microporous hydrophobic membrane bioreactor, *J. Am. Oil Chem. Soc.*, **62**(6), 1016–1021.

HOQ, M.M., YAMANE, T., SHIMIDZU, S., FUNADA, T. and ISHIDA, S., 1995, *J. Am. Oil Chem. Soc.*, **62**(6), 1016–1021.

HORIUTI, Y. and IMAMURA, S., 1978, Stimulation of chromobacterium lipase activity and prevention of its adsorption to palmytol cellulose by hydrophobic binding of fatty acids, *J. Biochem.*, **83**, 1381–1385.

HYAKE, Y., OHKUBO, M. and TERAMOTO, M., 1991, Lipase-catalyzed hydrolysis of rinaphthyl esters in biphase system, *Biotechnol. Bioeng.*, **38**, 30–36.

ISONO, Y., NABETANI, H. and NAKAJIMA, M., 1995a, Wax ester synthesis in a membrane reactor with lipase-surfactant complex in hexane, *J. Am. Oil Chem. Soc.*, **72**(8), 887–890.

ISONO, Y., NABETANI, H. and NAKAJIMA, M., 1995b, Lipase surfactant complex as catalyst of interesterification and esterification in organic media, *J. Ferment. Bioeng.*, **80**(2), 170–175.

JENSEN, R.G., GALLUZZO, D.R. and BUSH, V.J., 1990, *Biocatalysis*, **3**, 307–316.

KANG, S.T. and RHEE, J.S., 1989, Characteristics of immobilized lipase-catalyzed hydrolysis of olive oil of high concentration in reverse phase system, *Biotechnol. Bioeng.*, **33**, 1469–1476.

KIAM, A., BHAVE, R.R. and SIRKAR, K.K., 1984, Solvent extraction with immobilized interfaces in a microporous hydrophobic membrane, *J. Membr. Sci.*, **20**, 125–145.

KLIBANOV, A.M., 1990, Asymmetric transformations catalyzed by enzymes in organic solvents, *Acc. Chem. Res.*, **23**, 114–120.

KLIBANOV, A.M. and ZAKS, A., 1985, Enzyme-catalyzed processes in organic solvents, *Proc. Natl Acad. Sci., USA*, **82**, 3192–3196.

KODERA, Y., NISHIMURA, H., MATSUSHIMA, A., HIROTO, M. and INADA, Y., 1994, Lipase made active in hydrophobic media by coupling with polyethylene glycol, *J. Am. Oil Chem. Soc.*, **71**(3), 335–338.

LANGRAND, G., SECCHI, M., BUONO, G., BARATTI, S. and TRYANTOPHYLIDES, C., 1985, *Tetrahedron Lett.*, **26**, 1857–1860.

MALCATA, F.X., HILL, C.G. and AMUDSON, C.H., 1991, Use of a lipase immobilized in a membrane reactor to hydrolyze the glycerides of butter oil, *Biotechnol. Bioeng.*, **38**, 853–868.

MALCATA, F.X., GARCIA, H.S., HILL, C.G. and AMUDSON, C.H., 1992a, Hydrolysis of butteroil by immobilized lipase using a hollow-fibre reactor: Part I. Lipase adsorption studies, *Biotechnol. Bioeng.*, **39**, 647–657.

MALCATA, F.X., REYES, H.R., GARCIA, H.S., HILL, C.G. and AMUDSON, P.H., JR., 1992b, Kinetics and mechanisms of reactions catalyzed by immobilized lipases, *Enzyme Microb. Technol.*, **14**, 426–446.

MARTINS, J.F., DA PONTE, M.N. and BARREIRAS, S., 1993, Lipase catalyzed esterification of glycidol in organic solvents, *Biotechnol. Bioeng.*, **42**, 465–468.

MELONVILLE, F.X., LOPEZ, J.L. and WALD, S.A., 1990, in AMBRAMOWICZ, D.A. (ed.) *Biocatalysis*, pp. 167–177, van Nonstrand Reinhold, New York.

MOLINARI, R., DRIOLI, E. and BARBIERI, G., 1988, Membrane reactor in fatty acid production, *J. Membr. Sci.*, **36**, 525–534.

MOLINARI, R., SANTORO, M.E. and DRIOLI, E., 1994, Study and comparison of two enzyme membrane reactors for fatty acids and glycerol production, *Ind. Eng. Chem. Res.*, **33**(11), 2591–2599.

MOMSEN, W.E. and BROCKMAN, H.L., 1981, The adsorption to and hydrolysis of 1,3-didecanoyl glycerol monolayers by pancreate lipase. Effects of substrate packing density, *J. Biol. Chem.*, **256**, 6913–6916.

MUSHANTA, A., FORSSELL, P. and PONTANEN, K., 1993, Application of immobilized lipases to transterification and esterification reactions in non-aqueous systems, *Enzyme Microb,. Technol.*, **15**, 133–139.

NATOLI, M., 1994, *Membrane catalitiche ad alta specifiatà*. Ph.D. thesis, University of Calabria.

NATOLI, M., NICOLOSI, G. and PIATTELLI, M., 1990, Enzyme catalyzed alcoholysis of flavone acetates in organic solvent, *Tetrahedron Lett.*, **31**, 7371–7374.

OKUMURA, S., INAI, M. and TSUJISAKA, Y., 1980, *J. Biochem.*, **87**, 205–211.

PRASAD, R. and BHAVE, R.R., 1986, Further studies on solvent extraction with immobilized interfaces on a microporous hydrophobic membrane, *J. Membr. Sci.*, **26**, 79–97.

PRASAD, R. and SIRKAR, K.K., 1992, Membrane based solvent extraction, in WINSTON, W.S. and SIRKAR, K.K. (eds) *Membrane Handbook*, pp. 727–763, van Nonstrand Reinhold, New York.

PRONK, W., KERKHOF, P.J.A.M., VAN HELDEN, C. and VAN'T RIET, K., 1988, The hydrolysis of triglycerides by immobilized lipase in a hydrophilic membrane reactor, *Biotechnol. Bioeng.*, **32**, 512–518.

RASTRUP-NIELSEN, T., PEDERSEN, L.S. and VILLADSEN, J., 1990, *J. Chem. Tech. Biotechnol.*, **48**, 467–482.

REYES, H.R. and HILL, C.J., JR., 1994, Kinetic modelling of interesterification reactions catalyzed by immobilized lipase, *Biotechnol. Bioeng.*, **43**, 171–182.

RUCKA, M. and TURKIEWICZ, B., 1989, *Biotechnol. Lett.*, **11**, 167–172.

RUCKA, M., TURKIEWICZ, B., ZUK, J.S., KRYSTYNOWICZ, A. and GALAS, E., 1991, Hydrolysis of plant oils by means of lipase from *Rhizopus nigricans*, *Bioprocess Eng.*, **7**, 133–135.

SHAW, J.F., CHANG, R.C., WANG, F.F. and WANG, Y.J., 1990, Lipolytic activities of a lipase immobilized on six selected supporting materials, *Biotechnol. Bioeng.*, **35**, 132–137.

SONNET, P.E., 1987, *J. Org. Chem.*, **52**, 3477–3481.

SONNET, P.E., 1988, *J. Am. Oil Chem. Soc.*, **65**, 900–904.

SUGIURA, M. and ISOBE, M., 1974, *Biochim. Biophys. Acta*, **34**, 195–200.

SUGIURA, M. and ISOBE, M., 1975, *Chem. Pharm. Bull.*, **23**, 1226–1230.

SUGIURA, M., ISOBE, M., MUROYA, N. and YAMAGUCHI, T., 1974, *Agric. Biol. Chem.*, **38**, 947–952.

SUGIURA, M., OIKAMA, T., HIZANO, K. and INUKAI, T., 1977, *Biochim. Biophys. Acta*, **488**, 353–358.

TANIGAKI, H., SAKATA, M. and WADA, H., 1993, Hydrolysis of soybean oil by lipase with a bioreactor having two different membranes, *J. Ferment. Bioeng.*, **75**(1), 53–57.

TSUJISAKA, Y., OKOMURA, S. and IWAI, M., 1977, *Biochim. Biophys. Acta*, **48**, 415–422.

VAIDYA, A.M., BELL, G. and HALLING, P.J., 1992, Aqueous-organic membrane bioreactors. Part I. A guide for membrane selection, *J. Membr. Sci.*, **71**, 139–149.

VALIVETY, R.H., JONNSTON, G.A., SUCKLING, C.J. and HALLING, P.J., 1991, Solvent effects on biocatalysis in organic systems: equilibrium position and rates of lipase catalyzed esterification, *Biotechnol. Bioeng.*, **38**, 1137–1143.

VERGER, R., 1980, Enzyme kinetics of lypolysis, In PURICH, D.L. (ed.) *Methods in Enzymology*, Vol. 64, pp. 340–392, Academic Press, New York.

WEST, J.B. and WANG, C.-H., 1987, *Tetrahedron Lett.*, **28**, 1629.

YAMANE, T., ICHIRYU, T., NAGATA, M., VENO, A. and SHIMIDZU, S., 1990, *Biotechnol. Lett.*, **36**, 1063–1069.

YANG, D. and RHEE, J.S., 1992, Continuous hydrolysis of olive oil by immobilized lipase in organic solvent, *Biotechnol. Bioeng.*, **40**, 748–752.

ZAKS, A. and KLIBANOV, A.M., 1984, Enzymatic catalysis in organic media at 100 °C, *Science*, **224**, 1249–1251.

ZAKS, A. and KLIBANOV, A.M., 1988, Enzymatic catalysis in non-aqueous solvents, *J. Biol. Chem.*, **263**, 3194–3201.

7

Membranes and Membrane Reactors in Artificial Organs

L. DE BARTOLO AND E. DRIOLI

Research Institute on Membranes and Modelling of Chemical Reactors, IRMERC-CNR and Department of Chemical and Materials Engineering, University of Calabria, I-87036 Arcavacata di Rende (CS) Italy

7.1 Introduction

Polymeric materials have become widely used as components of medical devices and implants, drug delivery systems, diagnostic assays, bioreactors and bioseparation processes. Most of the recent research activities in biomedical engineering have focused on organ replacement and extracorporeal treatment. Organ or tissue failure is a serious problem in the health care sector (Langer and Vacanti, 1993). In dollars, the cost of treatment, support and loss of productivity for people who suffer tissue loss or organ failure is estimated in the United States to exceed $400 billion per year. Furthermore, therapeutic treatment or organ transplantation are not always possible: for example, in the United States, fewer than 3000 donor livers are available annually while 30 000 patients die from liver failure (Hubbell and Langer, 1995). Since 1943, when Kolff and Beck in Holland first used an artificial kidney for the removal of catabolites from patient's blood, membranes have been used in medical devices (Drioli and Catapano, 1990). Membrane processes are used effectively for the intra- and extracorporeal treatment of patients with various pathologies for the removal of endogenous or exogenous toxins from blood (plasmapheresis, hemodialysis; hemodiafiltration) or for gas exchange with blood (blood oxygenation). The adoption of membranes is also extremely attractive with systems where biocatalysts in the form of mammalian cells, enzymes or tissue fragments are used. In fact, membranes of suitable molecular mass cut-off are used in bioartificial pancreas, in bioartificial liver using isolated cells, and as selective barriers to prevent the immune system components from coming into contact with implants while allowing nutrients and metabolites to permeate freely to and from cells.

Membranes are used in therapeutic blood treatment because they act as selective barriers between two phases: the blood and a physiological solution or oxygen-enriched stream.

In medical devices, membranes exhibit in general the same properties they present in microfiltration, ultrafiltration and dialysis processes. The pore sizes of microfiltration membranes range from 10 to 0.05 μm. Ultrafiltration is typically used to retain macromolecules from a solution, the lower limit being solutes with molecular mass of a few thousand daltons. Ultrafiltration and microfiltration membranes, as discussed in Chapter 1, are considered as porous membranes in which rejection is determined mainly by the size and shape of the solutes relative to the pore size of the membrane and in which the transport of chemical species is determined by a transmembrane pressure gradient:

$$J_v = K \Delta P \qquad (7.1)$$

where

J_v = volume flux $[LT^{-1}]$
K = permeability constant
P = hydrodynamic pressure $[ML^{-1}T^{-2}]$

This transport equation, with the different pore geometries of membranes that imply different models, may be described in a first approximation as (Mulder, 1991)

$$J = \frac{\varepsilon r^3}{8\eta\tau} \frac{\Delta P}{l} \qquad (7.2a)$$

in the case of parallel cylindrical pores; or as

$$J = \frac{\varepsilon^2}{K\eta S^2(1-\varepsilon)^2\tau} \frac{\Delta P}{l} \qquad (7.2b)$$

in the case of pores resembling close-packed spheres, where

ε = porosity
η = viscosity $[ML^{-1}T^{-1}]$
τ = tortuosity
l = membrane thickness $[L]$
S = surface area $[L^2]$

In dialysis, solutes diffuse from one side of the membrane to the other side according to their concentration gradients. The transport can be described by a diffusion equation:

$$J_s = \frac{D_s K_s}{l} \Delta C_s$$

where

C_s = solute concentration $[ML^{-3}]$
D_s = solute diffusion coefficient $[L^2T^{-1}]$
J_s = solute flux $[ML^{-2}T^{-1}]$

As a result, separation between solutes is obtained as a difference in diffusion rates across the membrane arising from differences in molecular size. High flux can be obtained using membranes as thin as possible. Considering that the diffusion coefficient is dependent on molecular mass of the chemical species, the dialysis process is more efficient than ultrafiltration processes in terms of separation between chemical species with more disparate molecular masses (Table 7.1).

Membrane operations exploiting the properties of molecular separation of species of interest, particularly in the blood, have been very successful in recent years and still today represent one of the most widespread applications of membranes. The potential of membrane bioreactors for the production of new artificial organs or improving the performance of the traditional ones is appreciated in the case of the artificial pancreas and artificial liver.

Table 7.1 Membrane processes in biomedical treatments

Membrane structure	Microfiltration	Ultrafiltration	Dialysis
	(A)Symmetric porous	Asymmetric porous	Homogeneous
Material	Hydrophilic-hydrophobic polymers (e.g. cellulose, polypropylene)	Hydrophilic polymers (e.g. polysulfone, polyacrylonitrile)	Hydrophilic polymers (e.g. cellulose, Cuprophan)
Thickness	~10–150 µm	~150 µm	~10–100 µm
Pore size	~0.05–10 µm	~1–100 nm	
Driving force	Pressure (<2 bar)	Pressure (1–10 bar)	Concentration differences
Separation principle	Particle size	Particle size	Difference in diffusion rate, solution diffusion

In this chapter the various biomedical applications will be discussed and reviewed. Artificial kidneys, plasma therapy equipment and artificial lungs are basically separation devices. However, their performances offer important indicators in the development of other artificial organs based instead on membrane reactors.

7.2 Artificial Kidney

7.2.1 *Hemodialysis*

In hemodialysis, membranes are used as artificial kidneys for people suffering from renal failure, for the purpose of removing toxins from patient's blood as urea, creatinine, uric acid and other compounds that are usually eliminated by the kidneys through the formation of urine. Today, worldwide about 600 000 dialysis patients use more than 60 million dialyzers per year (Shaldon and Vienken, 1996). These numbers are increasing annually by 6%.

The membrane function in a dialyzer simulates that of the nephron. Nephrons comprise glomerules and tubules. The first phase of urine formation is the filtration of plasma through the glomerule. The glomerular ultrafiltrate is absorbed through the tubule; 99% of water and important electrolytes are absorbed through diffusion mechanisms and active transport. In this way, the exogenous and endogenous toxins are concentrated and eliminated in the urine. The glomerular ultrafiltrate is produced with a flow rate of 125 ml min^{-1} (Eckert and Randall, 1982). In hemodialysis the patient is connected to an extracorporeal circuit by a veno-arterial shunt. Arterial blood goes through the dialyzer with a flow rate of approximately 200 ml min^{-1}, while buffer solution goes through the dialyzer in the opposite direction on the other side of the membrane with a flow rate of approximately 500 ml min^{-1} (Catapano and Drioli, 1994). Separation of metabolic toxins from the patient's blood is realized by differences in diffusion rates across the membrane (Figure 7.1). Table 7.2 lists some compounds of the glomerular filtrate in relation to molecular mass and concentration between ultrafiltrate and filtrate. The cut-off of dialysis membranes is lower than that of the glomerule. Chemical species with molecular masses

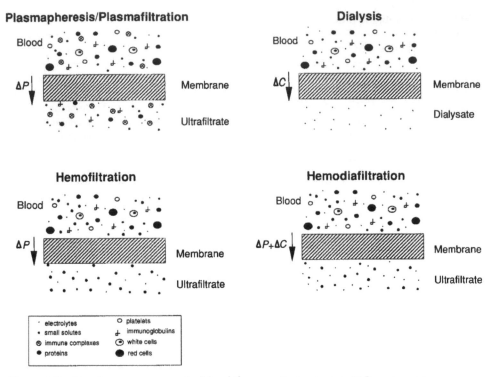

Figure 7.1 Membrane processes in blood therapeutic treatment. (ΔP = pressure difference; ΔC = concentration difference.)

Table 7.2 Compounds of the glomerular ultrafiltrate (Eckert and Randall, 1982)

Solute	Molecular mass (Da)	Radius[a] (nm)	Size[b] (nm)	[Ultrafiltrate]/[plasma]
Water	18	0.11	·	1
Urea	60	0.16	∘	1
Glucose	180	0.36	○	1
Saccharose	342	0.44	○	1
Insulin	5 500	1.48	○	0.98
Myoglobin	17 000	1.95	○	5.4	0.75
Hemoglobin	68 000	3.25	○	5.4	0.03
Serum albumin	69 000	3.55	○	15.0	<0.01

[a] Radius calculated from diffusion coefficient.
[b] Size calculated from X-ray diffraction.

between 5000 and 12 000 daltons, which are eliminated by natural kidneys, are not removed by hemodialysis treatment and their accumulation might be responsible for some complications of hemodialysis (anaemia, bone and joint pain, neuropathy, itching). For these reasons, new developments in dialysis membranes are devoted to realizing a 'high-flux dialyzer' and preparing highly permeable membranes such as poly(methyl methacrylate) (BK-F) (Bonomini et al., 1996) that, compared with the conventional PMMA and cellulose acetate, have demonstrated satisfactory dialytic removal of solutes including β_2-microglobulin, which plays an important role in complications of dialysis failure (Gurland et al., 1986).

7.2.2 Hemofiltration

To improve the assistance to uremic patients, new techniques based on the principle of hemofiltration have been introduced for removing toxins from patients' blood. The separation properties of membranes used in hemofilters are closer to those of the renal glomerule; membranes have a range from 40 to 50 kDa. In this way, toxins of intermediate molecular mass can be eliminated, avoiding their accumulation in the blood and consequent complications (Figure 7.1). In hemofiltration, excess water can be removed and, owing to convective mass transfer, solute transport is independent of the molecular size, in contrast to the diffusion process dominant in conventional dialysis, which favours elimination of small solutes. In fact, uremic middle-sized molecules are more effectively eliminated during hemofiltration than during dialysis. In the hemofiltration process, the large amount of lost filtrate containing important solutes needs to be replaced with an equal volume of substitution fluid containing those solutes. On the other hand, it is recognized that in hemofiltration the use of membranes with higher cut-off involves a loss of selectivity of the membrane: small solutes are removed less effectively than during dialysis, thus limiting the reduction of treatment time. To overcome this disadvantage, a new dialysis method was developed combining both high rate of ultrafiltration and efficient diffusion: hemodiafiltration. Owing to combination of convective mass transfer and diffusion, the clearance values of both small and larger molecules are significantly higher than in hemofiltration and hemodialysis alone with the same membrane.

7.3 Plasma Therapy

7.3.1 Plasmapheresis

Plasma therapy can be regarded as the most important treatment in some immunological and dysmetabolic diseases due to accumulation of macromolecules such as immunoglobulin G (IgG), IgM, cryoglobulins, immune complexes and LDL cholesterol (Siami and Stone, 1993). Patients with polyneuropathy, neuropathy, systemic glomerulonephritis and familial hypercholesterolemia are treated with plasmafilters for the removal of specific factors. Microfiltration membranes (e.g. ethylene vinyl alcohol (EVAL), polypropylene, polyethylene) with pore sizes ranging from 0.1 to 0.8 µm and membrane surface area of 0.2–0.5 m^2 are used. The high cut-off of the membranes means that only the cellular component of blood is rejected from membrane; thus, proteins and solutes important for osmotic equilibrium and for the coagulation system are eliminated along with toxins, as shown in Figure 7.1. The lost fluid is continuously substituted with frozen plasma or physiological solution containing albumin. The operative conditions are generally chosen to obtain a flow rate of ultrafiltrate between 10 and 60 ml min^{-1}. Treatment involves plasma exchange two or three times (Catapano and Drioli, 1994).

7.3.2 *Plasma Filtration/Adsorption*

Unlike plasma exchange, plasma filtration requires little or no albumin replacement. This procedure consists of the filtration of plasma through membrane filters with different pore sizes to remove toxins and to retain clotting factors and proteins (Figure 7.1). Smaller pore size plasmafilters are used to remove molecules with molecular mass $\geq$ 180 kDa such as IgG, immune complexes and cryoglobulins. Larger pore filters remove macro-molecules with molecular mass $\geq$ 900 kDa such as IgM and LDL cholesterol (Cooney, 1980). The efficacy of plasmafiltration in the removal of toxins is comparable with that of plasma exchange, but the requirement for albumin is greatly reduced. This procedure is used in patients with Guillain-Barré syndrome, chronic idiopathic demyelinating poly-neuropathy. The use of plasmafilters in the treatment of these patients permitted effective removal of pathogenic high molecular mass substances.

Plasma treatment with sorbents is an alternative technique used for the removal of toxins from patients' blood. This technique consists of the separation of plasma from whole blood by filtration through microporous membranes, the removal of toxins from the plasma through sorption, and the reinfusion of the purified plasma into the blood through filters that prevent the passage of the sorbent. Multisorbent systems contain anion exchange resin and activated charcoal for the removal of intermediate molecular mass solutes and molecules.

7.4 Artificial Lung

Blood oxygenation by membrane devices is used safety and effectively in cardiovascular and pulmonary surgery. Nowadays the extracorporeal membrane oxygenation system is an accepted technique for temporary lung support. Blood from systemic veins (e.g. vena cava) flows by pumping to a membrane device for oxygenation, and the oxygenated blood returns to the aortic branch in the systemic circulation. The membrane in devices for blood oxygenation has to have an extensive surface area (2–10 m^2) for gas exchange between blood and the oxygen-enriched stream. The membranes used in oxygenators are generally hydrophobic, permeable only to gas and not to liquids (e.g. polypropylene, Teflon, silicone). In this way, the resistance to gas transport through the membrane is reduced and consequently also the surface area of membrane necessary to obtain suffi-cient transmembrane gas flux.

The performance of membrane oxygenators closely approximates that of the natural lung, although in the natural lung the gas exchange surface (50–100 m^2) is an order of magnitude greater than in the artificial lung. Consequently, to saturate the blood a higher residence time in the device and a more oxygen-enriched gas current are necessary (Catapano and Drioli, 1994).

Most existing membrane devices adopting hollow-fibre membranes are operated in 'inside flow' mode. Blood flows inside the membrane lumen, and gas flows outside the hollow-fibre membranes in the shell in countercurrent. The rate of mass transfer is deter-mined by the transmembrane driving force and by the serial resistance to mass transport of the membrane and of the two boundary layers at the membrane interfaces with blood and with the enriching stream. A new type of membrane module has been developed to reduce blood boundary layer resistance to mass transport, by flowing the blood outside regularly spaced hollow fibres: this is the 'extraluminal flow device'. Membrane mats are prepared of hollow fibres knitted together, at an angle with respect to the threads. In this

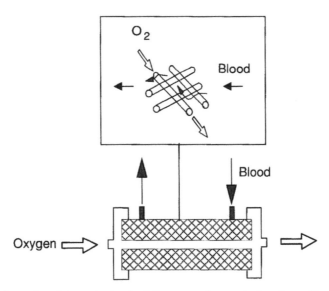

Figure 7.2 Schematic of the extraluminal flow membrane device (from Catapano et al., Blood Flow Outside Regularly Spaced Hollow Fibres: the Future Concept of Membrane Devices? *Int. J. of Artificial Organs*, 1992, 15(4): 327–30).

module, blood flows through the membrane mesh, and gas is fed countercurrently into the membrane lumen. The membrane arrangement induces mixing in every section of the membrane module to an extent that depends on membrane angle with respect to the main direction of blood flow (Figure 7.2) (Catapano et al., 1992).

Other methods for extrapulmonary augmentation of blood gas transfer have recently been developed utilizing intracorporeal rather than extracorporeal perfusion, such as the intravenacaval membrane oxygenator (IVOX). This system should be safer than extracorporeal perfusion and is accompanied by fewer and less severe complications. However, the quantity of extrapulmonary gas transfer accomplished by IVOX is usually less than that achieved by high-flow extracorporeal perfusion (Mortensen, 1993).

7.5 Bioartificial Pancreas

In patients with insulin-dependent diabetes, the cells that normally produce insulin do not function, so patients must receive insulin injections to regulate their blood glucose level. In the long term this results in so-called 'diabetic complications' such as kidney failure and neuropathy, which lead to the severe incapacity or death of the patient. Transplantation of the whole pancreas, pancreatic tissue fragments or islets of Langerhans would ensure the required glucose-related insulin secretion, but the rejection of implants and scarce availability of donor organs limits this therapeutic approach. An alternative approach is the development of a membrane bioartificial pancreas using isolated islets of Langerhans or single beta-cells segregated by means of membranes. Since 1970 when W.L. Chick and colleagues transplanted isolated islets, protected by a hollow-fibre ultrafiltration membrane (an acrylonitrile–vinyl chloride copolymer) with cut-off of 50 kDa, into dogs made diabetic by surgical removal of the pancreas, research efforts have been devoted to the development of a hybrid bioartificial membrane pancreas (HBMP) (Figure 7.3) (Chick et al., 1975).

The estimated number of islets required for therapeutic impact is about 700 000 per patient. The critical issues for bioartificial pancreas concern immunoprotection, maintenance of cell viability and functions, reduction of diffusion resistance for nutrients and metabolites, and lastly glucose/insulin kinetics. The many different types of prosthetic devices proposed so far can be grouped into three main categories: extravascular devices, intravascular devices, and microencapsulated islets of Langerhans (Goosen et al., 1985). In the first case, the tissue is enclosed between membranes in a flat sheet configuration (Figure 7.3a), or in the lumen of hollow-fibre membranes (Figure 7.3b), or in the shell of hollow-fibre membranes (Figure 7.3c) and then implanted into an extravascular site. These systems generally suffer from an intrinsically slow insulin response following changes of blood glucose concentration, limited by the purely diffusive mass transport and by the fibroblastic response of the host. In the case of microencapsulated islets, the membrane, in the form of alginate gel, is formed around the islets of Langerhans, thus obtaining microcapsules with diameters of 400 μm (Figure 7.3d). Microcapsules may make better implants than hollow fibres because they offer better conditions for diffusion of nutrients to the insulin-producing cells and of waste products from them. However, most biochemicals used in microcapsule manufacture affect the functionality of the transplanted tissue and elicit a localized tissue response (Langer and Vacanti, 1993).

Intravascular membrane devices are designed so that the membrane separates the graft directly from the bloodstream of the host. These devices suffer from blood clotting at the interface between blood and the synthetic material of the membranes or at the point of access, but they are extremely attractive in terms of flexibility of design and use. Additionally, the implantation site can be chosen on the basis of reducing the response time of the prosthesis following an increase of blood glucose concentration. *In vivo* experiments on diabetic animals show that the performance of an HBMP depends on the way the device is implanted (the performance of an HBMP implanted as an arteriovenous shunt is better than when the unit is implanted as an arterioarterial shunt) and on bioreactor dynamic behaviour. Starling flow in the reactor shell has been demonstrated to improve reactor performance by enhancing solute mass transport and by reducing insulin bio-feedback to the immobilized islets of Langerhans (Catapano et al., 1990).

7.6   Bioartificial Liver

Liver failure results from infection, drugs, or as a part of the multiple organ failure syndrome. Since fulminant liver failure is potentially reversible, the extracorporeal bridging of liver function would be beneficial until the patient's own liver resumes functional activity. Over the past 30 years, a variety of supportive therapies for patients with acute liver failure have been proposed. Detoxification-based methodologies for liver support such as dialysis, hemofiltration and hemoperfusion have been proven ineffective because physical methods are not sufficient for the management of severe biochemical disorders. A bioartificial liver support device using isolated hepatocytes might constitute an effective therapeutic approach for the temporary treatment of patients with acute liver failure. Several studies indicate that hepatocytes are capable of supporting all essential hepatic functions and may supply biologically active substances that promote regeneration and repair of the damaged liver being supported.

Two different artificial liver supports are proposed: implantable systems including microcarrier-attached hepatocytes, spheroid aggregate hepatocytes or microencapsulated hepatocytes, and extracorporeal devices including membrane devices, as illustrated in

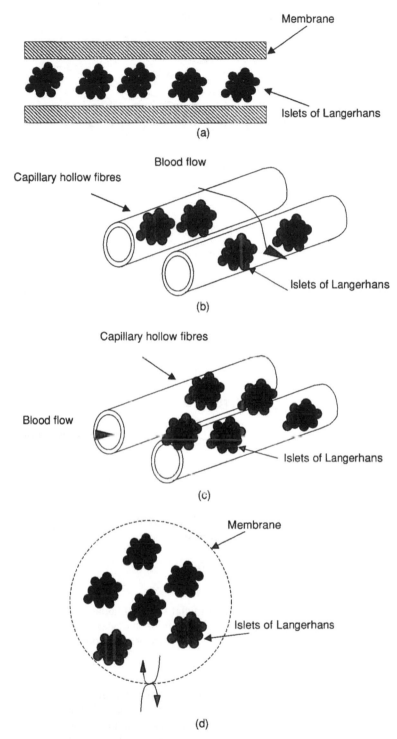

Figure 7.3 Bioartificial pancreas devices: (a) islets of Langerhans loaded between flat-sheet membranes; (b) islets of Langerhans loaded inside hollow fibres; (c) islets of Langerhans loaded outside hollow fibres in the extracapillary compartment; (d) microencapsulated islets of Langerhans.

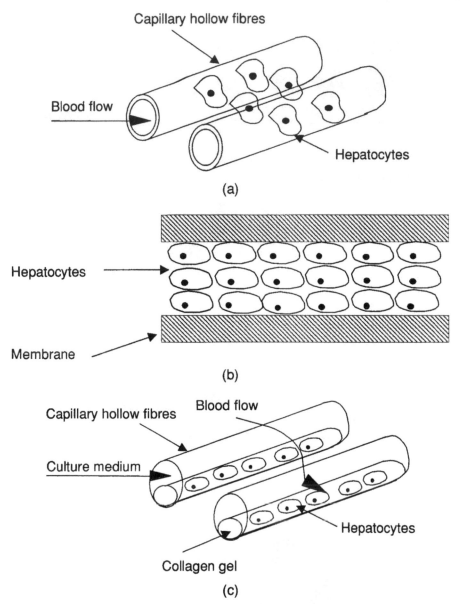

Figure 7.4 Different configuration of bioartificial liver: (a) hepatocytes loaded outside hollow fibres in the extracapillary compartment; (b) hepatocytes loaded between flat-sheet membranes; (c) hepatocytes entrapped in a three-dimensional contracted gel matrix inside hollow fibres.

Figure 7.4. The first system is untested in clinical trials as it involves immunosuppressive therapy and the support materials may inhibit the reticuloendothelial system. In the extra-corporeal membrane device, the membrane acts as an immunological barrier between the patient's blood and isolated hepatocytes and also as the substratum for cell adhesion, thus improving the metabolism of anchorage-dependent cells such as hepatocytes. The first clinical report of a bioartificial membrane liver was released in 1987 (Matsumura et al., 1987). This device consisted of a hepatocyte suspension that was separated from the

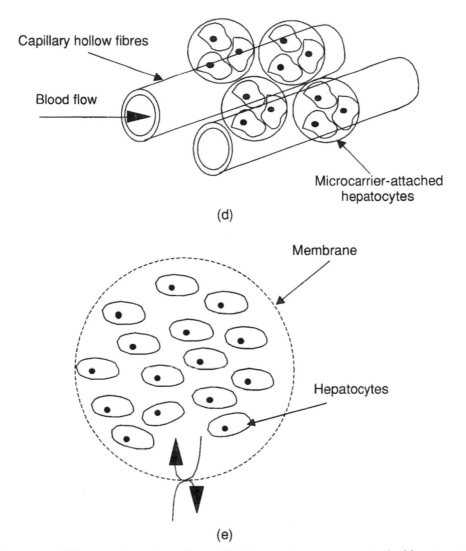

Capillary hollow fibres

Blood flow

Microcarrier-attached
hepatocytes

(d)

Membrane

Hepatocytes

(e)

Figure 7.4   Different configuration of bioartificial liver: (d) microcarrier-attached hepatocytes in the extracapillary compartment; (e) bioartificial liver using microencapsulated hepatocytes.

patient's blood by a cellulose acetate dialysis membrane. Different extracorporeal devices using hollow-fibre membranes or flat-sheet membranes have since been proposed. In devices using hollow-fibre membranes, isolated hepatocytes are usually loaded outside of the hollow fibres in the extracapillary compartment, while blood, plasma, or culture medium flows through the lumen of the hollow fibres (Figure 7.4a). In devices using flat-sheet membranes, hepatocytes are loaded between flat membranes in a sandwich fashion, while blood or culture medium flows outside of the membranes as shown in Figure 7.4b. Cells may be free in suspension, attached to walls, or attached to microcarriers (Figure 7.4d).

A key issue concerning the development of a bioartificial liver is the maintenance of the long-term viability and functions of hepatocytes: oxygen transport resistance and catabolite accumulation may limit hepatocyte viability and metabolism (Catapano et al.,

1996a, 1996b). New culture models are in development to address this problem. A hollow-fibre bioreactor using hepatocytes cultured in the hollow-fibre lumens has been proposed (Nyberg et al., 1993). In this device hepatocytes are entrapped in a three-dimensional gel matrix that is contracted to increase the perfusion of cells so as to improve the supply of oxygen and nutrients and the removal of catabolites (Figure 7.4c). More recently there has been developed a modular flat-sheet bioreactor using porcine hepatocytes cultured within extracellular matrix between oxygen-permeable flat-sheet membranes. This device consists of a multitude of stackable flat membrane modules, each having an oxygenating surface area of 1150 cm^2. Isolated liver cells are located at distances of 10–$20 \text{ }\mu\text{m}$ in the extracellular matrix. This bioreactor provides culture conditions that improve liver-specific functions and assist proliferation of liver cells (Bader et al., 1997). To overcome limitations in the cultivation of primary hepatocytes, Sussmann and Kelly (1993) proposed the use of an immortalized liver cell line as the biological element in the extracorporeal device. Sussmann used a cloned human cell line derived from a hepatoblastoma and selected for liver-specific functions. This device consisted of 10 000 individual hollow fibres with cut-off of 70 kDa. It has been tested on 11 patients with fulminant hepatic failure: four patients were sustained until transplantation, two survived without the need for transplantation, and five died of complications. A disadvantage of this unit is the potential for seeding tumour cells into patients. Mass transfer limitations to and from the extracapillary space along with bioreactor scale-up pose potential limitations for hollow-fibre designs. Furthermore, in membrane hybrid liver support devices, to meet the cells' metabolic oxygen requirements about 15–40% of cell mass must be in direct contact with the membrane; thus cell interaction with the surface of the membrane used for cell culture needs to be carefully considered. It has previously been demonstrated that the physicochemical and morphological properties of membranes affect adhesion, viability and the kinetics of metabolic reactions of hepatocytes (Catapano et al., 1996c; De Bartolo et al., 1997). In the design of a membrane hybrid liver support device it is important to choose the optimal membrane and to take account of the effect of the membrane surface properties in the rate equations of the relevant metabolic reactions.

7.6.1   *Extracorporeal Blood Detoxifiers*

Hollow-fibre membrane reactors with covalently bound enzymes have been proposed as extracorporeal blood detoxifiers or as devices for reducing arginine and asparagine content in the blood of leukemic patients. A group of enzymes from pig liver cytosol, known as glutathione *S*-transferase, have been immobilized onto the active side of commercial hemodialyzers in order to reduce the toxin content in blood. These enzymes change the polarity of the toxins through conjugation to glutathione. The resulting shift from hydrophobic to hydrophilic nature facilitates release through the membrane wall. The arginine and asparagine content in the blood of leukemic patients has been recently related to growth of neoplastic cells; therefore their removal should hinder further cell development. Asparaginase and arginase have been immobilized in commercial Spiraflow hemofilters in order to support anti-leukemia therapies. *Ex vivo* and *in vivo* experiments circulating blood through these devices generally confirm an effective decrease in blood arginine and asparagine content after a relatively short time of extracorporeal circulation treatment (Rossi, 1981).

In *in vivo* experiments, long reactors may not attain maximum reaction rates, owing to high pressure drops arising from blood viscosity. Higher apparent reaction rates result

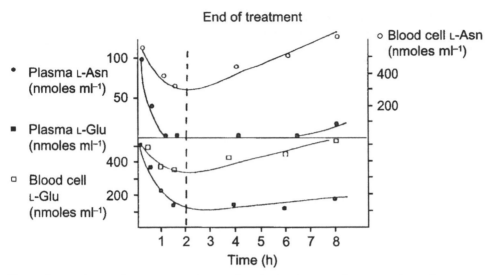

Figure 7.5 L-Asparagine concentration vs time in *in vivo* experiments performed on rats. The extracorporeal device is a 'tube-and-shell' membrane reactor in which asparaginase is bound to the outer surface of Cuprophan membranes (Mazzola and Vecchio, 1981).

when operating with shorter reactors at higher flow rates but under the same pressure drop. *In vivo* extracorporeal circulation experiments on rats with asparaginase bound to the outer surface of Cuprophan hollow membranes assembled in a tube-and-shell reactor configuration confirm that even this kind of reactor can eliminate asparagine after a short period of time, i.e. 3–4 hours (Figure 7.5). While asparagine depletion in plasma is accomplished quickly, in blood cells it proceeds more slowly. Unfortunately, after the treatment, the asparagine content in blood cells rises suddenly; when it levels off at its initial value, asparagine again appears in the plasma. This behaviour has been confirmed by experiments on leukemic patients. Even though the use of such reactors as extra-corporeal units can result in an effective decrease of asparagine or arginine content in blood, its effect is limited to a few hours after the treatment. This suggests that more than one metabolite has to be removed from blood at the same time to make the extracorporeal treatment effective (Mazzola and Vecchio, 1981).

Up to four enzymes involved in the metabolic pathways of purine bases — allan-toinase, allantoicase, uricase and catalase — have been immobilized together by means of glutaraldehyde on the outer surface of cellulosic hollow fibres.

Enzyme stability and activity can be enhanced by performing enzyme immobilization in the presence of inert proteins, such as albumin, or polyamines in order to shift the maximum enzyme activity towards blood pH.

7.7 Biocompatibility Characteristics of Membranes in Artificial Organs

Membranes used in medical devices for extracorporeal treatments must possess functional characteristics such as selective permeability, biostability and promotion of cell growth. In extracorporeal blood purification, membranes must possess appropriate permeability characteristics to enable controlled removal of solutes and water, as in hemodialysis, high-flux hemodialysis and hemofiltration, or the separation of blood cellular elements from plasma, as in membrane plasma separation. Blood exchange takes place in the artificial

Table 7.3 Commercial membrane materials used in artificial organs

Membrane material	Medical device[a]
Cellulose and derivated (esters, acetates, nitrates)	HD, HDF, HF, AL
Cuprophan	HD, HDF, AL
Cellophane	HD, HDF
Polysulfone	HD, HDF, HF, PF
Polycarbonate	HD, HDF, HF, AL
Poly(methyl methacrylate)	HD, HDF
Ethylvinylalcohol	HD, HDF, AP
Poly(vinyl acetate)	HD, HDF
Polyimide/poly(etherimide)	AP
Polyamide	AL
Polypropylene	Ox, PF, AL
Polytetrafluoroethylene	Ox
Silicone	Ox

(Adapted from Catapano and Drioli, 1994).
[a] HD, hemodialysis; HF, hemofiltration; HDF, hemodiafiltration; PF, plasmapheresis;
Ox, blood oxygenation; AP, artificial pancreas; AL, artificial liver.

lung, where membranes allow the addition of oxygen and removal of carbon dioxide at rates suitable for clinical application (Courtney et al., 1993). Thus, the choice of the membrane material for use in the extracorporeal device is dependent on its permeability characteristics as well as on its physicochemical properties in relation to the separation process. Table 7.3 lists some commercial membrane materials used in artificial organs.

In the case of hybrid bioartificial organs like pancreas and liver, isolated cells are compartmentalized in polymeric membranes that support a number of important functions for the success of these devices. Membranes should act as a barrier to immune components present in the patient's blood and should permit the rapid passage of key metabolites such as nutrients and oxygen from the surroundings to the cell compartment. In such devices, cells are in contact with semipermeable membranes and the surface properties of the membranes could affect the viability and metabolism of cells. Thus, one major approach has been to try to influence the extent and the character of the foreign-body response by modifying the surface composition and properties of the polymer.

7.7.1 Blood–Membrane Interactions and Cell–Membrane Interactions

In membrane bioartificial organs using isolated cells as the biological component, semipermeable membranes act as barriers between the patient's blood and the cells in order to avoid immunological attack by the host against the biological component of the device. Membranes of suitable molecular mass cut-off are permeable to nutrients and metabolites but reject the immunocompetent species present in the blood of the patient. The membranes may also act as the substrate for cell attachment and culture, as in the case of anchorage-dependent cells (i.e. hepatocytes), thus improving their viability and metabolic function. Figure 7.6 shows isolated hepatocytes in adhesion culture on a polymeric microporous membrane. After 22 hours of culture, isolated rat hepatocytes re-establish cell-to-cell contacts and form three-dimensional aggregates, as *in vivo*.

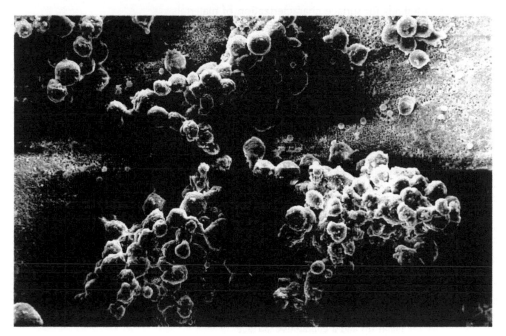

Figure 7.6 Isolated hepatocytes in adhesion culture on a polymeric microporous membrane, × 400.

In a membrane artificial organ, cells come into contact with the membrane surface. Therefore, the response of the biological components depends on the surface properties of the membrane used. Physicochemical properties including surface composition, surface charge, surface energy, and surface morphology have been shown to affect cell adhesion and behaviour. Surface properties may affect cell adhesion and metabolism by influencing the ability of the substratum to adsorb protein and/or by altering the conformation of the adsorbed protein of the extracellular matrix or by guiding cell adhesion on the basis of the surface topography. Continued advances are being made in surface characterization methods and a number of selective methods are available for material surface characterization, as reported in Table 7.4.

Four *in vitro* reactions of cells with surface have been postulated: (a) cells may interact strongly but not specifically with a surface, leading to attachment and de-differentiation; (b) cells may interact weakly with a nonadhesive surface — cells are not activated to perform specific functions; (c) cells may react strongly and specifically with a surface — cells attach and perform highly specific functions; (d) cells are encased in a three-dimensional porous matrices, permitting them to function in a physiological fashion. Literature data underline the importance of microporosities and microtextures present on the polymer surface in cell adhesion. Recently, studies performed on isolated hepatocytes cultured on semipermeable polymeric membranes have indicated that wettable and rougher surfaces enhance the adhesion and metabolism of isolated hepatocytes (Catapano et al., 1996c; De Bartolo et al., 1997). Microstructures (i.e. pores) on membrane surfaces offer attachment points for cell adhesion, improving cell viability.

Physicochemical properties also play a dominant role in the modulation of cell–membrane interactions, favouring the adsorption and consequent conformational change of important proteins such as fibronectin and vitronectin which are principal mediators of the cell adhesion.

Table 7.4 Methods for surface characterization of materials

Method	Principle	Depth analyzed	Spatial resolution	Analytical sensivity
Contact angle	Liquid wetting of surfaces is used to estimate the energy of surfaces	0.3–2 nm	1 mm	Low or high depending on the chemistry
ESCA	X-rays cause the emission of electrons of characteristic energy	1–25 nm	10–150 μm	0.1 at%
Auger electron spectroscopy	A focused electron beam causes the emission of Auger electrons	5–10 nm	10 nm	0.1 at%
SIMS	Ion bombardment leads to the emission of surface secondary ions	1 nm–1 μm	50 nm	High
FTIR-ATR	IR radiation is absorbed in exciting molecular vibration	1–5 μm	5 μm	1 mol%
STM	Measurement of the quantum tunnelling current between a metal tip and a conductive surface	0.5 nm	0.1 nm	Single atoms
AFM	Measurement of the attractive or repulsive force between atoms in a surface and in a tip	0.5 nm	0.3 nm	Single atoms
SEM	Secondary electron emission caused by a focused electron beam is measured and spatially imaged	0.5 nm	4 nm tipically	High, but not quantitative

Ratner (1993).
ESCA, electron spectroscopy for chemical analysis; SIMS, secondary ion mass spectroscopy; FTIR-ATR, Fourier transform infrared spectroscopy; STM, scanning tunnelling microscopy; ATM, atomic force spectroscopy; SEM, scanning electron microscopy.

However, other forces are also involved in the cell–polymer surface interaction such as physical forces and specific receptor to surface protein ligand interactions. To engineer better membranes for cell culture it is of considerable importance to characterize the single attributes of surface properties in cell interaction, and to define the relationship between such properties and cell behaviour.

7.8 References

BADER, A., DE BARTOLO, L. and HAVERICH, A., 1997, Initial evaluation of the performance of a scaled-up flat membrane bioreactor (FMB) with pig liver cells, In G. CREPALDI, A.A. DEMETRIOU, M. MURACA (eds) *Bioartificial Liver: the Critical Issues*, pp. 36–41, CIC International Editions.

BONOMINI, M., FIEDERLING, B., BUCCIARELLI, T., MANFRINI, V., DI ILIO, C. and ALBERTAZZI, A., A, 1996, New polymethylmethacrylate membrane for dialysis, *Int. J. Artif. Organs*, **19**(4), 232–239.

CATAPANO, G. and DRIOLI, E., 1994, Membrane polimeriche per trattamenti extra-corporei del sangue, *La Chimica e l'Industria*, **76**, 106–115.

CATAPANO, G., IORIO, G., DRIOLI, E., LOMBARDI, C.P., CRUCITTI, F., DOGLIETTO, G.B. and BELLANTONE, R., 1990, Theoretical and experimental analysis of a hybrid bioartificial membrane pancreas: a distributed parameter model taking into account starling fluxes, *J. Membr. Sci.*, **52**, 351–378.

CATAPANO, G., WODETZKI, A. and BAURMEISTER, U., 1992, Blood flow outside regularly spaced hollow fibres: the future concept of membrane devices?, *Int. J. Artif. Organs*, **15**(4), 327–330.

CATAPANO, G., DE BARTOLO, L., LOMBARDI, C.P. and DRIOLI, E., 1996a, The effect of oxygen transport resistances on the viability and functions of isolated rat hepatocytes, *Int. J. Artif. Organs*, **19**(1), 61–71.

CATAPANO, G., DE BARTOLO, L., LOMBARDI, C.P. and DRIOLI, E., 1996b, The effect of catabolite concentrations on the viability and functions of isolated rat hepatocytes, *Int. J. Artif. Organs*, **19**(4), 245–250.

CATAPANO, G., DE BARTOLO, L. and DRIOLI, E., 1996c, Effect of membrane surface morphology on the kinetics of ammonia elimination and oxygen consumption by rat hepatocytes, *Int. J. Artif. Organs*, **19**(9), 544.

CHICK, W.L., LIKE, A.A. and LAURIS, V., 1975, Cell culture on synthetic capillaries: an artificial endocrine pancreas, *Science*, **187**, 847–856.

COONEY, D.O., 1980, *Biomedical Engineering Principles*, Marcel Dekker, New York.

COURTNEY, J.M., IRVINE, L., JONES, C., MOSA, S.M., ROBERTSON, L.M. and SRIVASTAVA, S., 1993, Biomaterials in medicine: a bioengineering perspective, *Int. J. Artif. Organs*, **16**(3), 165–171.

DE BARTOLO, L., CATAPANO, G., DELLA VOLPE, C. and MIGLIARESI, C., 1997, Role of the material surface properties in liver cell–membrane interaction, *Artif. Organs*, **21**(6), 526.

DRIOLI, E. and CATAPANO, G., 1990, Membrane in processi di trattamento del sangue. In V. CAMBI (ed.), *Trattato Italiano di Dialisi*, cap.4, pp. 1–25, Wichtig, Milan.

ECKERT, R. and RANDALL, D., 1982, *Fisiologia Animale*, Zanichelli, Milan.

GOOSEN, M.F.A., O'SHEA, G.M., GHARAPETIAN, H.M., COHN, S. and SUN, A.M., 1985, Optimization of microencapsulation parameters: semipermeable microcapsules as a bioartificial pancreas, *Biotechnol. Bioeng.* **27**, 146.

GURLAND, H.L., LYSAGHT, M.J. and SAMTLEBEN, W., 1986, Immunomodulation: clinical aspects, *Artif. Organs*, **10**(2), 122–127.

HUBBELL, J.A. and LANGER, R., 1995, *Tissue engineering C&N*, American Chemical Society, 42–53.

LANGER, R. and VACANTI, J.P., 1993, *Tissue Engineering*, Science, **260**, 920.

MATSUMURA, K.N., GUEVARA, G.R., HUSTON, H., HAMILTON, W.L., RIKIMARU, M., YAMASAKI, G. and MATSUMURA, 1987, M.S. hybrid bioartificial liver in hepatic failure: preliminary clinical report, *Surgery*, **101**(1), 99–103.

MAZZOLA, G. and VECCHIO, G., 1981, *Hollow Fibre and Capillary Membranes in New Separation Processes*, pp. 52–58, CNR, Rome.

MORTENSEN, J.D., 1993, Extrapulmonary blood gas exchange: why? who? when? how?, *Artif. Organs*, **17**(6), 510.

MULDER, M., 1991, *Basic Principles of Membrane Technology*, Kluwer, Dordrecht.

NYBERG, S.L., SHATFORD, R.A., PESHWA, M.V., WHITE, J.G., CERRA, F.B. and HU, W-S, 1993, Evaluation of a hepatocytes entrapment hollow fibre bioreactor: a potential bioartificial liver, *Biotechnol Bioeng.*, **41**, 194–203.

RATNER, B.D., 1993, New ideas in biomaterials science — a path to engineered biomaterials, *J. Biomed. Mater. Res.*, **27**, 837.

ROSSI, V., 1981, *Int. J. Artif. Organs*, **4**, 102–107.

SHALDON, S. and VIENKEN, J., 1996, Biocompatibility: is it a relevant consideration for today's hemodialysis?, *Int. J. Artif. Organs*, **19**(4), 201–214.

SIAMI, G.A. and STONE, W.J., 1993, Efficacy and safety of plasmapheresis using plasmafilters, *Artif. Organs*, **17**(6), 538.

SUSSMANN, N.L. and KELLY, J.H., 1993, Improved liver function following treatment with an extracorporeal liver assist device, *Artif. Organs*, **17**(1), 27–30.

Index